FOOD AND SUSTAINABILITY IN THE TWENTY-FIRST CENTURY

THE ANTHROPOLOGY OF FOOD AND NUTRITION

Series Editor: *Helen Macbeth*

Eating is something all humans must do to survive, but it is more than a biological necessity. Producing food, foraging, distributing, shopping, cooking and, of course, eating itself are all deeply inscribed as cultural acts.

This series brings together the broad range of perspectives on human food, encompassing social, cultural and nutritional aspects of food habits, beliefs, choices and technologies in different regions and societies, past and present. Each volume features cross-disciplinary and international perspectives on the topic of its title. This multidisciplinary approach is particularly relevant to the study of food-related issues in the contemporary world.

FOOD AND SUSTAINABILITY IN THE TWENTY-FIRST CENTURY

Cross-Disciplinary Perspectives

Edited by
Paul Collinson, Iain Young, Lucy Antal and Helen Macbeth

berghahn
NEW YORK · OXFORD
www.berghahnbooks.com

First published in 2019 by
Berghahn Books
www.BerghahnBooks.com

Library of Congress Cataloging-in-Publication Data
A catalog record for this book is available from the Library of Congress
Library of Congress Cataloging in Publication Control Number:
2019008427

British Library Cataloguing in Publication Data
A catalogue record for this book is available from the British Library

ISBN 978-1-78920-237-3 hardback
ISBN 978-1-80073-916-1 paperback
ISBN 978-1-78920-238-0 ebook

https://doi.org/10.3167/9781789202373

CONTENTS

ILLUSTRATIONS

Figures

Tables

PREFACE

....

Following in the tradition of the Berghahn Books series, *The Anthropology of Food and Nutrition*, this volume provides a cross-disciplinary selection of chapters focusing on a subject of great contemporary interest in relation to food: sustainability. Existing material on food and sustainability is spread widely over the academic literature of different disciplines, as well as featuring prominently in the reports and other outputs of non-governmental organisations, aid agencies, political groups and national and international governmental bodies. It is also an issue of significant interest in the media and popular discourse. Yet, there is insufficient consensus about what 'sustainability' actually means, probably exacerbated by its widespread use in so many different contexts. This volume brings together some approaches to the very important and contemporary topic of food and sustainability from different perspectives, with a commentary on the use of the word, *sustainability*, in the Introduction.

The editors are fully aware that no book can encompass all the perspectives on this important and very current but broad subject. However, because of the urgency of understanding issues associated with food and sustainability in our contemporary world, we feel that every attempt to bring together social and biological, academic and practitioner approaches to the topic is important. Whereas the chapters are, of course, very different, the identification and development of linkages between them show the importance of adopting a holistic approach to the study of food and sustainability.

The editors and contributors to this volume benefited from multidisciplinary discussions held at the University of Liverpool, UK. The following organisations from Liverpool and the nearby areas of Northern England contributed in different ways to the development of those discussions: The Institute of Risk and Uncertainty at the University of Liverpool, Liverpool Food People, Green Guild of Students, Merseyside Recycling and Waste Authority, the Co-operative Group UK, Princes Foods, Quorn (Marlow Foods), Typhoo Teas, Ibis Hotel, Homebaked Anfield, Constellations, The

Granby4Streets Community Land Trust, Real Junk Food Liverpool and The Grapes Community Garden. We are grateful to Diana Roberts for assistance on the Index, and finally, the editors would like to thank all the contributors to this volume and the publishers, Berghahn Books.

Paul Collinson
Iain Young
Lucy Antal
Helen Macbeth

INTRODUCTION
FOOD AND SUSTAINABILITY IN THE TWENTY-FIRST CENTURY

....

Paul Collinson, Iain Young, Lucy Antal and Helen Macbeth

Complex Definitions

Food is a subject that can be analysed from a multitude of different disciplinary perspectives, from biochemical research on nutrients and neurology of taste to social observations about the concepts, meanings and beliefs of different cultural groups regarding specific food items and their preparations – and many areas in between. Some studies are limited to a few individuals (perhaps for medical purposes); others are on one social, cultural or ethnic group, or comparisons between a few such groups; and others pursue a national, international or global focus. There are, of course, many further areas of consideration, both academic and practical, such as those concerned with agriculture, the environment, industry and marketing, and more. For this reason, the International Commission on the Anthropology of Food (ICAF) has consciously fostered cross-cultural and cross-disciplinary discourse in its publications. That is the case for this volume. Human food does indeed provide nutrition, but its diversity and distribution are determined by all the factors involved in its availability in each situation, and its consumption is affected by the acceptability of food options and its preparation both at individual and cultural levels. Even though what is considered edible or eatable (MacClancy et al. 2007) varies significantly between different communities, it is not hard to find agreement that human food means food that humans consume.

In contrast, there is less agreement about the meaning of sustainability. Diverse interpretations of the concept are based first on what variable is or is not being considered for sustaining, and secondly, on what level of that variable should be aimed at to qualify it as being sustained. As Peveri (this volume) points out, use of the word *sustainable* 'is at the same time ubiquitous and fuzzy'. Whereas the Latin basis of the word *sustain* means 'to hold up from below', still relevant today when referring to load-bearing structures, the word in English has expanded to encompass several other meanings. One use of the word relates to maintenance and continuation, but one must define clearly what variable is to be maintained. Is it the global or a local environment? In regard to food, is it the supply of food alone or perhaps of food and water, and to whom? Is it the diversity of species that provides human food or all biodiversity? Is some quality of life of humans being considered for their very survival and, if the latter, is it survival of all humanity or of just some human groups or even individuals? The point is that to maintain a given variable in an ever-changing ecosystem, other circumstances might well need to be changed to achieve that sustainability; thus, maintenance of one variable frequently depends on changes to others.

Another use of the word relates to the supply of necessities, nourishment and so on, and in this sense food sustainability may be equated with food security. However, food insecurity may mean different things in different circumstances; it may mean risk of starvation and death, or fluctuating periods of hunger, or uncertainty about the availability of some desirable or desired nutrients in the options for a varied diet. Discussions of food security are frequently linked with aspirations for equitable access to food, whether between individuals or groups, which thereby can convey a political and economic aspect. For many people, therefore, to achieve food sustainability also implies some change in political structures and policies, as well as in environmental conditions, rather than a continuation of current systems. So, whereas the concept of sustainability is about the maintenance of some variable, it is nearly always associated with change in other variables.

Drawing from these points, we posit a broad, but working, definition of food sustainability:

> Food sustainability for humans involves the ability, sustained over time, to produce or procure enough food to meet an individual's or a population's nutritional requirements, using production, distribution and disposal systems which have a neutral or beneficial impact on the environment and ecosystems, and that ideally are underpinned by forms of social justice that can ensure equitable access to food.

Numerous aspects of research and discussion are indicated by this definition. Whereas the chapters in this book span different disciplines and demonstrate diversity of interest, they nevertheless cover only a small proportion of this broad topic. The intention of this volume is to stimulate a cross-disciplinary dialogue concerning this contemporary and urgent issue.

Framework of this Volume

The framework on which the order of chapters in this volume has been arranged has been influenced by the five phases in food-related behaviour suggested by Jack Goody, who wrote:

> The study of the process of providing and transforming food covers the four main phases of production, distribution, preparation and consumption, ... to which can be added a fifth phase, often forgotten, disposal. (Goody 1982)

Since Goody's seminal work, his analysis of processes and phases has been referred to so often that it has become a fundamental analytical tool in anthropological discussions on food. In this volume, an introductory section consisting of four chapters provides a scene-setting overview of different aspects of sustainability. Goody's five phases of production, distribution, preparation, consumption and disposal are then used as a way of providing an organising structure for the remaining chapters, with each relating primarily to one of the phases. The editors are fully aware that any multidisciplinary focus inevitably leads to great diversity in subject matter and style, but employing Goody's phases as an organising framework allows linkages to be identified and explored. The exposure of the relationships between such different approaches demonstrates the importance of bringing together in one volume contributions from specialists in very different areas of study.

Chapter 1 discusses the theme of food and sustainability and introduces several different disciplinary perspectives on the topic. The necessity of a cross-disciplinary approach is strongly supported in the chapter, as the complexity of the interacting factors and processes involved is emphasised. Although the first paragraph envisages many perspectives on the topic, our overview of a few of them is of necessity limited, identifying a selection of approaches primarily from the human sciences.

Chapter 2 provides an overview of food insecurity in sub-Saharan Africa at the present time, exploring its different causes and drivers. Paul Collinson highlights issues such as population growth, economic and political issues, environmental stress, conflict and displacement, drawing out their relevance to discussions of sustainability. He argues that conflict is a major cause of food insecurity in a number of countries in sub-Saharan Africa, and that conflict resolution will be key to moving towards food sustainability in the future. Yet, he concludes that the latter remains a distant goal in this region at the present time.

Chapter 3 concerns a discussion of how the health of humans and of the environment can be integrated into ideas about sustainability as related to food, in a review of the construction and transformation of the concept of a 'Mediterranean Diet'. Xavier Medina describes how an earlier emphasis solely on health, originating in the 1960s (Keys 1970), has been transformed to include ideas surrounding sustainability, which he identifies as an adaptation rather than an eclipse of the original focus on health. The author discusses the diversity within the concept of the 'Mediterranean Diet' in relation to its acceptance as part of the UNESCO Intangible Cultural Heritages of Humanity, but his argument stresses that the transformation of the concept to include environmental elements and sustainability reflects wider cultural changes within the region.

In Chapter 4, Giovanni Orlando uses examples from his fieldwork in Palermo, Italy to illustrate global themes. He starts by referring to a Worldwatch Report (Worldwatch 2013) that queries whether achieving a sustainable society is possible, but he questions Leach's (2013) challenge to this, arguing from a post-Pasteurian point of view. He offers ethnographic perspectives based on his fieldwork regarding the consumption of organic foods and attitudes to these, and he reveals that a culture of (quiet) sustainability based on anxieties concerning pollution and environmental degradation is being created. Material from interviews reported in this chapter, as well as detailed comments about some globally significant pesticides, are integrated into his discussion of the theory of planetary boundaries, the Anthropocene and the concept of 'quiet sustainability' (Smith et al. 2015). This chapter exemplifies how ethnographic research in one area can be used as an analytical lens to highlight macro themes.

As explained above, the remaining chapters are organised according to Goody's five phases. Chapters 5, 6 and 7 are concerned with the resources to achieve food sustainability in a changing environment and can, therefore, be related to the process of production. The three chapters illustrate quite different perspectives on this. In Chapter 5, Martin Tena and colleagues focus on the vegetation in one Mexican canyon, the Barranca del Río Santiago (BRS), the contemporary environmental pressures on the area and the effects on food options for local people. They argue that for the land to produce enough food for food security in the future, attention must be paid to environmental sustainability through biodiversity. The chapter stresses the importance of the loss of phytogenetic resources in the area of this barranca, the loss of genetic biodiversity for food generally, and the urgency to search for new food resources. The authors point out that Mexico now ranks high on the International Union for Nature Conservation's list for number of endangered species in a country and it has more endangered species than any other country in Latin America. Tena et al. emphasise that much still needs to be learned regarding biodiversity in the region in order to understand food sustainability.

A more positive aspiration for sustainability is given by Iain Young in Chapter 6, who introduces a way forward for increasing food sustainability by explaining the opportunities offered by aquaponics – the production of plants irrigated by water fertilised by the nutrient-rich waste from tanks in which fish have been farmed. In turn, the plants use the nutrients, and the clean water can then be recycled back into the fish tanks. This process demonstrates the value of recycling in a closed circular system, providing opportunities for such production within and around urban areas. The chapter reports on some successful aquaponic projects and then stresses the importance of education about these processes, identifying some workshops about aquaponics and urban farming around the Liverpool area of the UK. Arguing that aquaponics is unlikely to replace traditional agriculture for easily transported products, Young nevertheless highlights its potential contribution to food sustainability in urban and peri-urban areas.

Carrying on the theme of increasing affordable food production in areas where people live, Helen Macbeth in Chapter 7 recalls successful food policies in the UK during the Second World War and asks if lessons can be learned from these. The emphasis of the chapter is on the allotments movement and people growing food for their families in their leisure time, encouraged by the dissemination of government-sponsored information regarding growing vegetables and using nutritious recipes. In wartime, vegetables were grown not only on the allotments but on any spare land, such as railway embankments and road verges. Interviews with some allotment holders today reveal that, for nearly all of them, the incentive is to produce healthy food and/or to engage in healthy physical activity; none of the respondents mentioned environmental sustainability, and only one mentioned cost. Whereas Macbeth's stress is on affordable sustainability of providing fresh vegetables for all members of society, there is an underlying hope linking to the cultural transformation, described by Xavier Medina (this volume), that would transform the perception of the value of growing vegetables for the family from just 'health' to 'health plus sustainability'.

Although the next two chapters describe very different situations, both are concerned with the cooperation between governmental and non-governmental organisations (NGOs), and others, in seeking to improve food distribution and food sustainability in deprived conditions. Chapter 8 has links with both chapters 6 and 7, as Lucy Antal introduces a project funded by the interaction of local government, NGOs and others to stimulate local food sustainability in six places in the UK, each facing 'significant challenges in relation to poverty, citizen health, environment and economy'. The Sustainable Food Cities Network (although not all six places are 'cities') was set up to develop a partnership in each place between local businesses, public agencies, NGOs and academic institutions. In her chapter, Antal introduces the scheme and reports on some of the results achieved in the different areas. This chapter exemplifies the importance of including the perspectives of those attempting to deliver food security in contemporary situations of need.

In Chapter 9, Peter Kaiser also discusses a situation of contemporary need, describing the lives of Karen refugees in camps on the Thai–Burmese border. In discussing the topic of food sustainability in relation to subsistence and the distribution of humanitarian aid, he points out that giving active attention to the protection of the environment, biodiversity, species welfare and natural resources, as well as the consumption of healthy food, is simply irrelevant to those for whom survival and day-to-day subsistence are all-absorbing concerns. Although some Karen in the camps were victims of forced migration, the long duration of these camps means that many have been born there. Kaiser makes clear that there has been no way that older generations have been able to pass on their traditional knowledge of cultivation, care of livestock and local foraging to younger generations, nor have there been any opportunities for farming or for education in modern, relevant technologies. Instead, dependency has developed in camp life. With the 2012 armistice, the Karen have been allowed to return to their former settlements, but without either traditional or new skills to ensure a self-reliant, sustainable lifestyle. Kaiser makes some practical suggestions for the management of refugee camps and recommends that our aims for sustainability globally do not deflect attention from the needs of people for whom survival is their daily priority.

Chapter 10 is concerned with food preparation. Valentina Peveri continues the themes highlighted in Chapter 9 in contrasting the concepts of sustainability current in the cosmopolitan societies of developed economies with a different form of 'frugal' sustainability pertaining to populations that have survived in adverse environmental conditions for generations. In presenting ethnographic information on the preparation of enset breads, porridge and gruel in south-western Ethiopia, Peveri identifies an essential point to make about sustainability, which is that the perennial need for frugality for survival in some societies has led to traditional foodways that have proved to be sustainable, allowing survival in adverse environmental conditions. Furthermore, she highlights the difference between the way in which enset is prepared and appreciated in the locale and the way in which the food is commonly perceived by international organisations. These tend to equate it with poverty and lack of development, while proffering development strategies based on a totally different, often cash-cropping, model associated with other crops. This chapter, while providing information on preparation, is also concerned with a whole food system based, in times of scarcity, on enset.

In relation to Goody's fourth phase, 'consumption', Michaël Bruckert in Chapter 11 also refers to discourses about sustainability emanating from and relating to 'well-fed' populations. His chapter, based on fieldwork in Tamil Nadu, is about meat consumption in India, which is still marginal even though a majority of the contemporary population would not consider themselves to be 'vegetarian'. He discusses the complex social, religious, political and economic pressures against meat eating, and the contemporary factors that are causing meat eating to rise and increasing the visibility of meat availability in urban areas. Then, with reference to the supply chain, he questions the sustainability of increased meat consumption on the subcontinent, especially

when the Indian government is encouraging an intensification of meat production. Bruckert considers this to be an 'Indian meat dilemma', in which the
benefits of higher meat consumption in terms of improved nutrition and food
security would only be achieved through animal factory farming and degradation of the environment. Insight into the complexity of pressures and issues
involved in creating food sustainability is central to this chapter.

Chapter 12 by Mabel Gracia-Arnaiz provides an interesting contrast to the
previous three chapters, but has several links with Chapter 8 in discussing the
complexity of factors due to contemporary economic pressures in Europe.
Her study of food insecurity for many people in Spain during the recent
economic crisis describes an alteration in food choices and consumption patterns. This chapter identifies the fact that people adopt a variety of strategies
to manage their needs, with new patterns of eating out (soup kitchens, street
food, etc.) and the search for cheaper foods having implications for cultural standards and family relationships. She points out that while these are
coping mechanisms, they also lead to social suffering, anxiety and stress and
therefore become a driver of political debate surrounding the concept of the
'right to food'. This compares in an interesting way with Orlando's chapter
(Chapter 4) and his ethnographic examination of the Palermo residents and
their fears about food pollution.

With regard to Goody's fifth phase, 'disposal', Nick Doran and Iain Young
in Chapter 13 take a broader perspective, as they examine the development of
food waste management strategies within the UK, highlighting the relevance
of research into this within institutions of higher education. They comment
on the challenges that exist, including the limited knowledge of the drivers
that give rise to food waste within the supply chain and the voluntary nature
of regulation governing food waste management in the UK. They introduce
current research into the specific considerations relating to the University of
Liverpool and the City of Liverpool.

However, waste has even greater importance in the discussion of food and
sustainability, as illustrated by the environmental organisation, Feedback[1],
which campaigns against unused food waste. Instead, as discussed further in
Chapter 1, Feedback supports the recycling of waste in a circular system that
includes low input. Chapter 14 offers an example of how sustainable waste
management combined with new technology can create a circular model
whereby disposal of food waste can be used to produce new food. Iain Young
describes the work of the BiFFiO project, which looks at the challenge of
waste produced by the aquaculture and agricultural industries, and suggests
solutions using new technology to produce energy and fertiliser.

Thus, our volume ends with a practical solution that points to hope for the
future, as well as telegraphing a crucial additional dimension which can be
added to Goody's five-phase model when addressing the topic of sustainability: this is the use of the disposal of waste in a circular system that returns the
waste to benefit production. We might call this new phase 'replenishment'.

Such circularity is commonly found in many traditional systems. However, in modern, industrialised societies the problems of waste disposal have significantly undermined environmental sustainability, and research into potential solutions for this are, therefore, becoming increasingly urgent.

Conclusion

Through the chapters in this book, we seek to demonstrate the importance of examining the issues of food and sustainability from a range of viewpoints and through cross-disciplinary discussion. Many of our contributors highlight some practical ways in which food sustainability can be achieved, both at local levels and on a more global scale. Building a resilient infrastructure, making human settlements inclusive, safe and sustainable, ensuring healthy lives and well-being for all, at all ages, and protecting and conserving our existing resources are as essential for the citizens of Palermo as they are for the Karen farmers or the populations of sub-Saharan Africa. Whereas all possible perspectives on food and sustainability cannot be encompassed within one volume, the diverse case studies within this book demonstrate the importance of creating cross-disciplinary academic and practitioner discussion about food and sustainability. Through this, we can learn from each other's experiences and work together to combat food insecurity, inequalities over food access and the effects of climate change by both adopting new technologies and learning from traditional farming practices, in order to create a circular food system that supports fauna, flora and people.

Paul Collinson
Oxford Brookes University, Oxford, UK

Iain Young
Institute of Clinical Sciences, University of Liverpool, UK

Lucy Antal
The Food Domain, Liverpool, UK

Helen Macbeth
Oxford Brookes University, Oxford, UK

Note

1. 'Feedback Global' website, published at https://feedbackglobal.org/about-us/. Last accessed on 30 November 2018.

References

Goody, J. (1982) *Cooking, Cuisine and Class: A Study in Comparative Sociology*, Cambridge University Press, Cambridge.

Keys, A. (1970) Coronary Heart Disease in Seven Countries, *Circulation*, 41(S1): 118–139.

Leach, M. (2013) Pathways to Sustainability: Building Political Strategies. In Worldwatch Institute *State of the World 2013: Is Sustainability Still Possible?* Island Press, Washington, DC, pp. 234–243.

MacClancy, J., Henry, C.J. and Macbeth, H.M. (eds) (2007) *Consuming the Inedible: Neglected Dimensions of Food Choice*, Berghahn Books, Oxford.

Smith, J., Kostelecký, T. and Jehlička, P. (2015) Quietly Does It: Questioning Assumptions about Class, Sustainability and Consumption, *Geoforum*, 67: 223–232.

Worldwatch Institute (2013) *State of the World 2013: Is Sustainability Still Possible?* Island Press, Washington, DC.

CHAPTER 1
TOWARDS A CROSS-DISCIPLINARY APPROACH TO FOOD AND SUSTAINABILITY IN THE TWENTY-FIRST CENTURY

....

Iain Young, Paul Collinson, Lucy Antal and Helen Macbeth

Introduction

In the previous chapter, the editors discussed the word 'sustainability'. We highlighted the difficulty of defining the concept, the importance of clarifying precisely the variable that is to be sustained, as well as emphasising the complexity of the interactions and interrelationships between different variables. There is a voluminous literature in which the variable to be sustained is the ecosystem within which food is produced; this may relate to climate and climate change, air purity, sufficient water, water purity, productivity of the soil and appropriate recycling of waste materials. Other research focuses on biodiversity as the variable to be sustained, either in one place, in a region or globally. Then there is the sustainability of the human activities needed to produce food. In many areas of study, the focus is on the sustainability of human populations themselves, frequently discussed under the heading of food security, where food has to be secured to sustain the population. The focus here can be on any combination of the social, environmental, economic and political factors that underpin food security, as well as its effects on the nutritional health and well-being of a population, differentiated perhaps by ethnicity, status, wealth or age. A key theme in much of the literature is how food production, distribution, preparation, consumption and disposal can be sustained for future generations in the face of such challenges as population growth, climate and environmental change, conflict and deepening social and geographical inequalities. In all cases the interaction of factors adds complexity, which is why we stress not just multidisciplinary perspectives, however incomplete, but the need to communicate between disciplines through a cross-disciplinary approach.

In this chapter, we introduce some of these variables and discuss the inter-relationships and interactions between them through our application of a multidisciplinary lens on human food and sustainability.

Global Perspectives

In recent decades, 'development' has come to be viewed as perhaps the primary variable to be sustained, judging by the widespread, almost ubiquitous use of the term 'sustainable development' both in academic and in political, practitioner and popular discourses. We, therefore, begin this chapter with the list of UN Sustainable Development Goals adopted at the United Nations Sustainable Development Summit in New York in 2015 (United Nations Department of Economic and Social Affairs 2015). These goals are intended as a 'blueprint to achieve a better and more sustainable future for all' by 2030 (United Nations 2018). Most of these goals interrelate in one way or another with the provision of food to humans – some more obviously than others.

'This compilation provides a summary of 17 initiatives – one for each of the goals …:

- Goal 1 – End poverty in all its forms everywhere
- Goal 2 – End hunger, achieve food security and improved nutrition, and promote sustainable agriculture
- Goal 3 – Ensure healthy lives and promote well-being for all at all ages
- Goal 4 – Ensure inclusive and equitable quality of education and promote lifelong learning opportunities for all
- Goal 5 – Achieve gender equality and empower all women and girls
- Goal 6 – Ensure availability and sustainable management of water and sanitation for all
- Goal 7 – Ensure access to affordable, reliable, sustainable and modern energy for all
- Goal 8 – Promote sustained, inclusive and sustainable economic growth, full and productive employment and decent work for all
- Goal 9 – Build resilient infrastructure, promote inclusive and sustainable industrialisation and foster innovation
- Goal 10 – Reduce inequality within and among countries
- Goal 11 – Make cities and human settlements inclusive, safe, resilient and sustainable
- Goal 12 – Ensure sustainable consumption and production patterns
- Goal 13 – Take urgent action to combat climate change and its impacts
- Goal 14 – Conserve and sustainably use the oceans, seas and marine resources for sustainable development

- Goal 15 – Protect, restore and promote sustainable use of terrestrial ecosystems, sustainably manage forests, combat desertification, halt and reverse land degradation and halt biodiversity loss
- Goal 16 – Promote peaceful and inclusive societies for sustainable development, provide access to justice for all and build effective, accountable and inclusive institutions at all levels
- Goal 17 – Strengthen the means of implementation and revitalise the global partnership for sustainable development.'
  (United Nations Department of Economic and Social Affairs 2015).

The dimensions envisaged by the United Nations (UN) for sustainable development are by no means the only perspectives on concepts of sustainability, but they provide many examples to consider. They also highlight the diversity of perspectives on this topic, which strongly supports the need for cross-disciplinary dialogue such as is set out in this volume.

Many would argue that the UN goals are utopian ideals and that even partially meeting these targets represents an enormous, perhaps even insurmountable, challenge for humanity. However, the achievements that have been recorded in improving food security, if not sustainability, worldwide over the past three decades are testament to the combined efforts of the international community, and they demonstrate what can be achieved through concerted multinational action. This is exemplified by the fact that the proportion of people suffering from undernourishment worldwide has declined from 18 percent in 1990 to 11 percent in 2016. The fall in the developing countries has been even more dramatic – from 23 percent to 12 percent (Food and Agriculture Organization et al. 2017: 7). This decline in the number of hungry people has occurred despite a rapidly expanding population in developing countries, which has more than doubled over the same period. However, there is no room for complacency, because undernourishment has risen since 2014, both in absolute terms (from 775 million people in 2014 to 815 million people in 2016) and as a proportion of the global population (from 10.7 percent in 2014 to 11 percent in 2016) (ibid.: 5). This last represents the first rise in the proportion of people going hungry since 2001–2 (ibid.: 5).

Although global food prices have been generally stable since the food price shocks of 2008–11, the cost of importing food is increasing, rising around 6 percent in 2017; these rises are even more marked in the least-developed countries (Food and Agriculture Organization 2017). Drought, flooding and protracted conflicts are all key factors in food insecurity. The United Nation's Food and Agriculture Organization Global Information and Early Warning System report 'Crop Prospects and Food Situation' (Food and Agriculture Organization 2018a) highlights conflict as a key driver for severe food insecurity (see also Collinson, this volume). This is a particularly important factor in parts of sub-Saharan Africa, the Middle East, South East Asia and elsewhere. This report also observes that conflict is a barrier to access to food, impedes aid and increases numbers of displaced people. Recent conflicts in Afghanistan, Iraq and Syria have brought chronic hunger to 7.6, 3.2 and 6.5 million people respectively. Unfavourable rainy seasons between 2015 and 2017 have driven food insecurity in Kenya, Ethiopia and Somalia, and drought in Mongolia has halved the country's wheat crop, whilst in the same period, Bangladesh has suffered three major episodes of flash flooding, destroying most of the country's rice crop.

With the population of the world set to increase from 8 to 10 billion between 2018 and 2050, it has been estimated that food production will need to increase by 70 percent over the same period to feed everyone on the planet (Food and Agriculture Organization 2009: 14). This seems to be an impossible challenge, except that, at the present time, there is more than enough food to provide nutrition for all. The Food and Agriculture Organization estimates that a third of all food grown across the world is wasted. The sustainable circular food model (outlined below) shows how a shift away from the linear food production model to a circular one would use fewer resources and repurpose surplus food, in the first instance, to feed people. Given that the improvements in food security which have been recorded over recent decades have occurred in parallel with a huge rise in the global population, there are some grounds for optimism for the future.

Nevertheless, we are facing a crisis in the ways that food is produced, distributed, consumed and disposed of, which has the potential to undermine the achievements that have been made (c.f. Devereux 2007). Food prices and supply remain unstable, fluctuating wildly in times of conflict and other global events (Piesse and Thirtle 2009). These fluctuations are likely to be exacerbated by climate change (Nelson et al. 2014) and by nutrient stripping in nutrient-poor soils in developing countries the world over (Jones et al. 2013). The disconnect between heavy nutrient use by arable farming and nutrient production in the form of animal waste from agriculture and human waste from urban concentration is discussed by Young (this volume).

In an insightful commentary on anthropological approaches to food and water security, Wutich and Brewis (2014: 445) suggest five factors that cause the risk of food insecurity: ecology, population, governance, markets and entitlements. In assessing the relevance of each, they argue that in relation to the first two, whilst environmental issues and population pressures may be significant factors, each is an insufficient explanation on its own. However, they consider the other three factors to have the potential to be the primary driver of food insecurity in most circumstances. In their demonstration of the relevance of anthropology and its sub-disciplines to understanding the issues surrounding sustainability, they stress the need for a cross-disciplinary approach which focuses on availability and entitlement to food.

A fundamental problem is differential access to food, both on a global level and within regions and countries (c.f. Sen 1981). If this can be addressed – and it is a big 'if' – it is at least a realistic possibility that a huge increase in food production will not be required. Whilst far too many people in the developing world do not have enough food to eat, many countries in the industrialised world are facing crises over what to do about surplus food and food waste. However, solving this problem on its own would do little to resolve food crises elsewhere. Global differentials in calories produced and consumed are partly a reflection of a food production and distribution paradigm which is fundamentally skewed in favour of the industrialised world (Biel 2016). Even within the developing and least-developed countries, food insecurity does not affect people equally: an absence of democracy, tensions between different social groups, neocolonialist production systems, violent conflict and different local environments, among other factors, lead to entrenched inequalities in food access. This is a problem that is increasingly affecting the industrialised world as well; witness the recent and contemporary rise of food banks in many European countries and the United States (c.f. Caplan 2016).

The situation is further complicated when we look more closely at the nutritional value of food. In many parts of the world, a simple, mutually exclusive distinction between undernutrition for the poor and overnutrition for the better off does not apply (Tanumihardjo et al. 2007; Delisle and Batal 2016). Undernutrition and obesity can exist in parallel. Research has demonstrated that as a country's income increases, obesity shifts from wealthier to poor groups (Bann et al. 2018), and patterns also start to emerge between obesity in adults and undernutrition in offspring. Further, undernutrition in early life predisposes the individual for obesity in adulthood.

Meanwhile, the environmental effects of current systems of food production, from the impacts on land use of the rise in cash cropping and monoculture, and the depletion of soils and soil nutrients in many areas of the world, to the pollution caused by transporting food over vast distances, are contributing to increasingly alarming changes in the global climate. These changes are likely to be difficult to halt, let alone reverse, over the coming decades – even if the political will exists to take action, which is by no means certain. The globalisation of the food supply chain has led to a situation where, increasingly, food production and redistribution are managed and 'owned' by a small number of shareholder-led companies, whose primary focus is profit (Taylor 2017).

It is clear, therefore, that the planet is currently facing a profound crisis not of food availability but of food sustainability. Intensive farming systems requiring high mechanisation and vast inputs of chemical nutrients, over-cultivation, soil erosion and inadequate nutrient replacement systems have led to both soil erosion and a precipitous decline in soil quality over recent decades, overwhelmingly concentrated in the developing and the least-developed parts of the world (Tan et al. 2005: 127–28). As well as contributing to food insecurity, this also leads to conflict between different communities as the amount of land for growing crops and grazing animals reduces.

A fundamental problem is the resources devoted to livestock production. Worldwide, land given over to animal grazing and the production of feed occupies over 80 percent of the world's agricultural land (Food and Agriculture Organization 2018b) but only contributes 18 percent of the total calories consumed (Poore and Nemecek 2018). Despite the efforts of the UN to reduce the amount of meat consumed (Henchion et al. 2017), global meat consumption is expected to increase by 76 percent by 2050 (WRAP 2018). The demand for grain and protein concentrates is closely related to meat production, with beef cattle consuming around 6 kg of plant protein for every kilogram of weight gained (Pimentel and Pimentel 2003). Thus, an increasing demand for meat accelerates the demand for agricultural crops. This has led to a call for the advancement of alternative protein sources such as cultured meat (Fayaz Bhat and Fayaz 2011), fungi (Asgar et al. 2010) and even insects (van Huis 2013).

The environmental costs of meat production, including increasing greenhouse gases through methane emissions, removing carbon sinks by turning forests over to grazing land and transporting live animals and meat around the world, are reaching a crisis point (Nguyen et al. 2012; EAT-Lancet Commission 2019). It has been estimated that in many parts of the developing and least-developed countries, 1,000 kilograms of carbon is released into the atmosphere for every kilogram of meat protein produced (Walsh 2013). In Europe, a recent report argued that the amount of land devoted to livestock production will need to be reduced by 60 percent by 2050 in order to comply with global greenhouse gas emission targets (Buckwell and Nadeu 2018: 9). Nor can we take any comfort in the hope that increasing amounts of carbon in the atmosphere will lead to increasing crop yields, since any small effects of this are more than offset by the deleterious effects of environmental change (Biel 2016: 68–69). The Intergovernmental Panel on Climate Change is unequivocal, concluding that: 'Based on many studies, covering a wide range of regions and crops, negative impacts of climate change on crop yields have been more common than positive impacts' (Intergovernmental Panel on Climate Change 2014: 7). A major three-year study published in 2019 by the Lancet medical journal concluded that the current global food system is the single most important contributor to environmental degradation and climate change, while unhealthy diets constitute 'a greater risk to morbidity and mortality than unsafe sex, alcohol, drug and tobacco use combined' (EAT-Lancet Commission 2019: 5). To move towards a more sustainable future, the report's authors recommended a fundamental overhaul of both food production systems and dietary habits, involving a radical reduction in red meat and dairy consumption combined with an unprecedented expansion in the cultivation of plant-based foods – changes that would have to be enforced by governments through taxes, prohibition of certain foods and rationing.[1]

Allied to these problems is the issue of food waste. It is estimated that a staggering 35–50 percent of all food produced in the world is not eaten, worth a total of one trillion US dollars (OLIO 2017). This problem largely affects industrialised societies where it also has severe environmental impacts, from the costs of producing the food in the first place (which accounts for a quarter of the world's freshwater), to the pollution caused by disposal of unwanted food. Changing consumer expectations in the industrialised world is also driving an exponential rise in food transport, whereby local variations in the seasonal availability of fruits and vegetables have been virtually eliminated in many countries as food is moved around to satisfy demand and the preferred 'just in time' hub and spoke model of the large-scale retailers (Fernie et al 2010). Increases in food imports have been paralleled by a squeeze on indigenous farmers, who are continually being forced by retailers to reduce costs in the most competitive markets.

The rise of 'fast food' in the West is an important driver of waste, creating a significant environmental problem associated with the disposal of not only the food itself but also its packaging. This marketing of fast food and other unhealthy, frequently cheap, processed foods to poorer people has created ingrained food cultures that perpetuate poor dietary choices. These are partly responsible for the increasing problem of obesity, which has become an epidemic in parts of Europe and the United States, and becomes relevant to overall environmental budgets when the cost of treating diseases associated with obesity places huge (and probably unsustainable) burdens on publicly funded health systems.

The United Nations development targets quoted above form a blueprint, which points to how the world can move towards a more sustainable future. However, for these targets to be achieved, even in part, concerted action will be required; it will need not only countries working together, but also citizens and public bodies operating in harness to create solutions to the problems that we face. This also highlights the potential importance of 'bottom-up' solutions to sustainability, which support ideals of democracy so that the interests of all are fairly represented in local and national fora.

Acting Local, Thinking Global

Awareness about environmental and climate change has spurred a huge global movement over recent decades, involving countless numbers of local activists across the industrialised and, increasingly, the non-industrialised worlds, who are striving to create a more sustainable future for their families and future generations. The following are a few representative examples:

- *Love Food, Hate Waste*[2] *(United Kingdom)*. Inspiring people to think about food waste and providing guidance about portioning and meal planning as well as recipes for 'leftovers'.

- *City of Austin's Zero Waste Initiative*[3] (Texas, United States). A city ordinance to require all restaurants to compost food waste – an initiative to reduce all landfills by 90 percent by 2040.
- *FUSIONS*[4] (Food Use for Social Innovation by Optimising Waste Prevention Strategies) *(European Union)*. After recognising that European Union food waste amounts to around 90 million tons a year, Brussels has pledged, through the FUSIONS programme, to reduce that number by half by the year 2025.
- *Last Minute Market*[5] (LMM) *(Italy)*. Working with farmers, processing centres, grocery stores, and other food sellers to reclaim food.
- The *Think, Eat, Save* campaign.[6] A partnership between UN Environment Programme, FAO and Messe Dusseldorf GmbH, which aggregates and shares information and methods of reducing food waste, including policy recommendations, and publishes inspirational projects on its website (http://www.thinkeatsave.org/).

In many countries, community groups, for example, the Real Junk Food Project,[7] are working to reduce food waste and improve justice surrounding access to food, often in partnership with local authorities. In the developing world, farmers and herders are organising collectively to introduce more sustainable food production systems (Herrero et al. 2010; Peacock and Sherman 2010; Thornton 2010), to help halt the degradation of their environments and to lobby governments and public authorities for changes in the way in which food production and distribution is organised. At the level of the family, recycling waste, including food waste, has become the norm in many industrialised countries. This has variable outcomes influenced by many factors, including age and gender; research has demonstrated that the elderly waste less than the young, and men waste less than women (Quested et al. 2013; Secondi et al. 2015; Jorissen et al. 2015; Melbye et al. 2017). There are also profound differences between countries.

Whilst the role of public policy and governmental action is obviously crucial, if there is to be a solution to global food sustainability, it is likely to be found as much in these local initiatives and changes in behaviour as it is in top-down approaches. Unfortunately, achieving a balance and synergy between the two, with public authorities working in harmony with communities and developing interventions, which are sympathetic to local circumstances and needs, has proved elusive in many areas of the world. In the case of developing countries, governments are too often part of the problem rather than the solution, leaving the solution in the hands of international agencies and NGOs – and international aid and development agencies have hardly had an unblemished record in this respect. Although the experience in industrialised countries is decidedly patchy as well, the situation has improved markedly over the past two decades or so, as governments increasingly recognise the importance of obtaining the buy-in of, and working in partnership with, target populations.

Some examples of the types of initiative which apply this model are provided later in the chapter.

All this suggests that research, which addresses how people are responding to the challenges they face through what might be termed 'the minutiae of food experiences', has a valuable role to play in informing public policy and bridging the gap between the local and the global. This is one of the aims of this book.

Anthropological Approaches

There is a burgeoning literature on the anthropology of sustainability, much of which focuses on food (for example, chapters in Lawrence et al. 2010; Murphy and McDonagh 2016; Brightman and Lewis 2017a). Some authors have argued that the crisis of sustainability and the evidence that the world is now entering a new geological and climatological epoch – the Anthropocene, driven by anthropogenic climate change – presents new opportunities for the discipline (for example, Guyot 2011; Brightman and Lewis 2017b; Latour 2017).

The key feature of anthropological studies is holism: recognising the complex interplay between different factors and drivers, and how they impact on the everyday experience of the human communities under scrutiny (c.f. Homewood 2017). The emic perspective is also prioritised. Rather than concentrating on large data sets and statistical analyses, anthropologists are more concerned with qualitative studies from the levels of the individual, the family or the community. Cross-cultural comparisons are then drawn to emphasise the similarities and, crucially, the differences pertaining in different social contexts. This approach, therefore, has significant value to practitioners, particularly in designing policies and programmes that are culturally sensitive and appropriate to particular environmental, social and political circumstances.

Anthropological approaches to food sustainability encompass a number of different themes:

- Experiences of and responses to food insecurity at the family and community level, drawing out cross-cultural comparisons of, for example, the localised impact of broad economic and social forces in driving food insecurity, and the coping strategies of different groups (e.g. de Waal 2005; Wutich and Brewis 2014; Carney 2015; Brückner and Cagler 2016; Collings et al. 2016).
- The relationship between everyday consumption habits and choices, and the wider social, cultural and political frameworks which both shape them, and are shaped by them. Specifically, this is a focus on how individual foodscapes relate to wider questions of sustainability and how concerns about sustainability are affecting patterns of consumption. Such an approach links strongly with environmental and

ecological anthropology (e.g. Abbots and Lavis 2013; Lockyer and Veteto 2013; Green and Asinjo 2015; Carolan 2015; Reuter 2015; Baer and Singer 2018).

- A focus on the moral and ethical dimensions of systems of food production and consumption. Such an approach may be highly political and orientated towards action anthropology, fostering change at local levels as a means of creating a more sustainable future for all (e.g. Clammer 2017).
- Studies emerging from development anthropology and the anthropology of development, and the nexus between the development and sustainability. Anthropologists have long taken a close interest in development issues and have played an important role as development practitioners for several decades. One of the discipline's key contributions has been to evaluate the neoliberal approach to development, and to act as advocates for populations targeted by development and humanitarian aid programmes (e.g. de Waal 1997; Nielsen 2014; Gupta and Pouw 2017).
- Biocultural and biological anthropological research, which examines the effects of resource scarcity or excess on human nutrition and health (e.g. Leatherman and Goodman 1997; Piperata 2008; Stevenson et al. 2012; Hadley and Wutich 2009).

Several of these themes are represented by different chapters in this volume.

Biological Approaches

As with the literature on the anthropology of food sustainability, climate change, increasing population and anthropogenic effects feature prominently in the literature of the biological sciences. Schmidhuber and Tubiello (2007) highlight that most climate simulation studies examine the direct effects of climate change on food production and availability through changes in agro-ecological conditions. Higher temperatures in temperate regions may increase food production as crop yields rise, but these benefits need to be set against the increasing frequency of severe weather events, such as hurricanes, floods and droughts. Schnidhuber and Tubiello also comment that many studies neglect social factors such as changing demand and food utilisation, and the increased prevalence of food-borne disease. This may start a vicious circle, with malnutrition increasing the susceptibility of a population to infectious disease, in turn reducing labour productivity and reducing food availability. Wheeler and von Braun (2013) also raise the issue of climate variability increasing risk and uncertainty in food supply, and call for mitigation actions towards a more resilient 'climate-smart food system'.

Anthropogenic effects are raised by Ziv et al. (2012) in explaining that the demand for electricity from a growing and increasingly affluent population around the Mekong River basin in China has promoted the building of hydroelectric dams. These dams will block fish migration paths and, in turn, impact fish populations in the biggest inland fishery in the world. Fisheries are a common topic in discussions of food security. Godfray et al. (2010) describe a threefold challenge: matching (1) changing demand from a larger and more affluent population with (2) environmentally and socially sustainable solutions, while (3) ensuring equitable access to food by the poor. Like the previous studies, they describe a multifaceted challenge which encompasses climate change, increasing urbanisation, technological challenges to increasing production and changing demand. This requires a similarly complex solution, involving changing agricultural practices, such as introducing large-scale farming operations to poor countries, GM crops with increased disease resistance and improved yield, and concerted global investment. They also highlight the benefits of waste reduction, both by improvements in the supply chain infrastructure, predominantly in poorer countries, and reductions in post-retail losses in the developed countries. For example, it is estimated that up to a third of the South Asia rice harvest is lost due to pests and spoilage after production, whilst in industrialised countries consumers have become so accustomed to cosmetically perfect produce that fruit and vegetables with even small blemishes are rejected.

Most biologists will be familiar with the concept of the increasing cost of maintaining predators at the top of a biomass (or energy) 'Eltonian' pyramid (Elton 1927). This describes the way in which a large number of producers support ever smaller groups of herbivores, small predators and predators of the predators. Like apex predators, the increasingly affluent global population demands more high-value meat and fish products. This drives an increase in global meat production (Henchion et al. 2014) whereby the farmed animals consume more protein than they produce. As we have noted above, this is profoundly unsustainable in terms of its impact on the environment and climate change.

The Need for a Circular System

It should be noted that our organisation of chapters along the lines of Goody's five phases suggests a more or less linear perception of the processes travelling in one direction, from production to consumption, through distribution and preparation, and ultimately, to disposal. Yet, all of these processes utilise resources and create waste, which, if not reused in some way, consumes further resources, including land, or becomes polluting. Such a system is not sustainable, as exemplified by the overproduction and unrecycled waste found in many wealthier parts of the world today.

Ideally, consumption and disposal should be intrinsically linked. This is described in the 'Food Recovery Hierarchy' pyramid (Figure 1.1), which demonstrates the actions that organisations and individuals could take to prevent and then divert food from becoming waste.

This pyramid starts at the top with the aim of reducing the amount of surplus food produced in the first place, and some of the onus for this must fall on the consumer as much as on the producer and retailer. A common argument made is that the public demands choice; therefore, wholesalers and retailers are merely responding to that demand and have no responsibility for consequent overproduction. Peveri (this volume) talks eloquently about the satisfaction to be had from a more frugal approach to food consumption, while Orlando (this volume) refers to worries from consumers about pollution caused by intensive food production. Does this mean that the power to change the system rests primarily with the consumer? Yet, just as retailers are influenced by what consumers buy, consumers are certainly influenced by retailers' marketing.

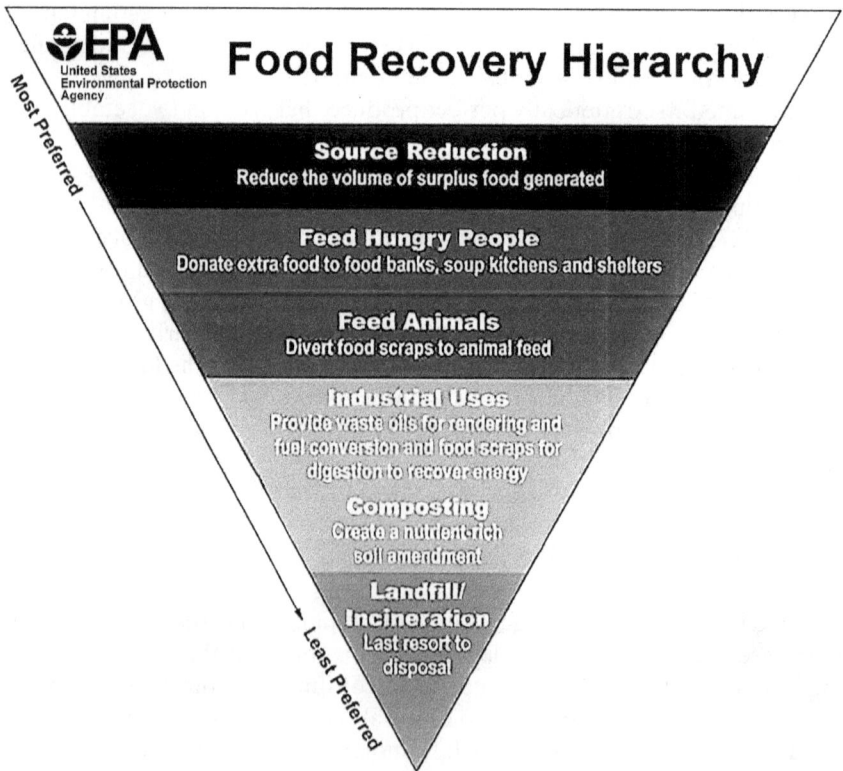

Figure 1.1 The Food Recovery Hierarchy pyramid. Figure created by the United States Environmental Protection Agency. Used with permission.

The next level of this food recovery hierarchy is concerned with feeding hungry people and food insecurity (see chapters by Antal, Collinson, Gracia-Arnaiz and Kaiser, this volume) and then with feeding animals. Young's chapter on aquaponics and Macbeth's on growing food on allotments link into the composting (feeding soils) layer of the diagram, which in this instance comes below industrial use (i.e. energy production). The final layer encompasses food that cannot be repurposed in any of the preceding ways, and thereby ends up in landfill or incineration. The pyramid can thus be inverted to demonstrate that the most preferable approach is to reduce overproduction. This can be considered as the first step in preventing food waste by not creating it in the first place. In the context of a linear progression from production to disposal, at least part of the solution must be to reduce waste at each stage and to utilise food in the most efficient way so that we produce, prepare and consume less or distribute more widely, and in turn, dispose of less. However, with any linear, unidirectional model, no matter how efficient, there is inevitably some waste created at each stage, which will always terminate in disposal. In contrast, a circular approach tries to rectify this, acknowledging that whilst overproduction still needs to be addressed, surplus or waste food can be repurposed in as sustainable a way as possible.

A circular food system, such as that illustrated in Figure 1.2, supports the aim of redirecting food waste in a holistic way to prevent as much as possible from entering into the final stage in the inverted pyramid, disposal through incineration or landfill. This also slightly reorders the Food Recovery hierarchy, as it shows a preference for feeding the soil over industrial repurposing (creating energy). 'Feedback', an environmental organisation that campaigns globally to end food waste at every level of the food system, describes food waste as 'a symptom of our broken food system' and has designed an alternative food system which moves away from the existing linear food system and towards a circular low-input one.

The circular food economy is very much driven by environmental concerns and advocates a smaller scale, more local approach to food system management that involves the consumer to a greater extent. Figure 1.2 shows a number of 'best use' loops that reuse food waste and give it a value. The nutrient recycling of food no longer suitable for human consumption can be converted to fish or animal feed; for example, 'The Pig Idea' is a campaign that encourages the feeding to pigs of food that still has nutritional value, but is no longer suitable for humans (Feedback 2018). Part of the overarching idea underpinning the circular food economy is that food waste should be treated as a resource, and as a resource it needs to be carefully husbanded within a smaller, more localised system, reducing the need for transportation and industrial processing. The majority of household food waste could be composted and managed locally, without the need for additional collection by vehicles. This is where a food strategy for a district or urban area can assist, by setting out goals and removing barriers through, for example, supporting citizens to change their behaviour through provision of a local composting site within walking distance.

So, in considering food sustainability, especially in industrial societies, it is clear that *circularity* is the bedrock on which true food sustainability can be built. If we are able to reuse what has previously been regarded as 'waste', we reduce the need for disposal – and thereby conserve energy and resources, including water and land. Linked to this ambition is the aim to reduce over-production so that a greater proportion of the food produced is consumed, rather than wasted and then disposed of. If, therefore, we are to move towards food sustainability, more calculated attention should be paid to 'quantities' in Goody's (1982) phases of production, distribution and consumption.

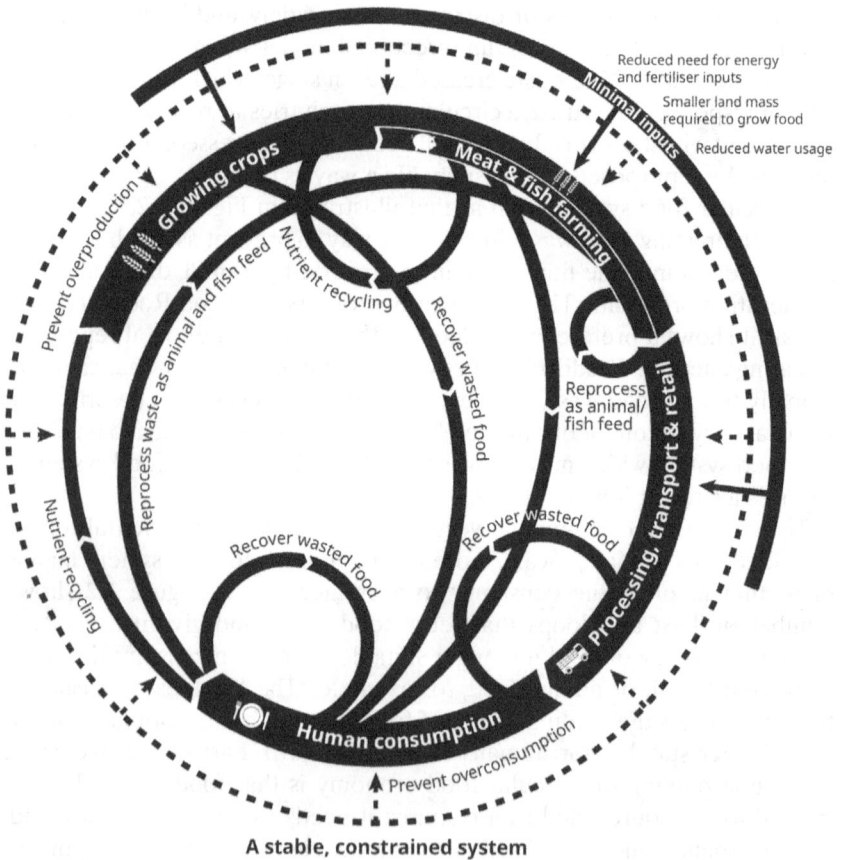

Figure 1.2 Diagram of a circular food system. Diagram created by the Feedback Global Organisation. Used with permission.

Furthermore, a sixth phase needs to be added, which we might term 'replenishment', the objective of which is to create a circular system whereby waste produced during each phase ultimately replenishes the inputs needed for previous phases.

An example of the circular system in action is provided by the Milan Urban Food Policy Pact (MUFPP), which was announced in February 2014 at the C40 Climate Change Leadership Group summit in Johannesburg. Conceived by the City of Milan as part of their EXPO 2015 theme 'Feeding the Planet, Energy for Life', it is an international protocol aimed at tackling food-related issues at an urban level, and is in the process of being adopted by other cities worldwide. To date, 165 cities, representing 450 million citizens, have signed the pact. It acknowledges that cities, which host over half of the world's population, have a strategic role to play in developing sustainable food systems (Biel 2016). It accepts that each city is different, but all are centres of economic, political and cultural innovation. A key agreed commitment, to which the mayors and representatives of local governments signed up, stated: 'We will encourage interdepartmental and cross-sector coordination at municipal and community levels, working to integrate urban food policy considerations into social, economic and environment policies, programmes and initiatives, such as, inter alia, food supply and distribution, social protection, nutrition, equity, food production, education, food safety and waste reduction' (Milan Urban Food Policy Pact 2015)

In 2015, the EU Commission agreed to finance the Food Smart Cities for Development project, fostered by the MUFPP and led by the City of Milan. A key objective was to raise public awareness about European food policies that focused on food security, sustainable development and decentralised cooperation (localism). Guidelines and a collection of good practices from partner cities and civil society organisations were published in 2016 (Milan Urban Food Policy Pact 2017). These drew from and were linked to the United Nations' Sustainable Development Goals, the United Nations' Ten-Year Framework on Sustainable Consumption and Production and the MUFPP, amongst other initiatives. The three main recommendations from the report are: to improve food governance and strengthen ties with society; to use Fair Trade as a way to encourage responsible and knowledgeable consumer behaviour; and to move towards decentralised cooperation that enables the tackling of global issues on a local level. Echoed in a number of chapters in this volume are the importance of local involvement and activity; food security as a driver for societal norms; and a consideration of how we, as individuals, must address our own consumption and disposal of food.

Projects such as the MUFPP point to what can be achieved with concerted public action. Its philosophy and aims are being replicated by millions of other grass-root initiatives across the industrialised and non-industrialised worlds, aimed at creating a more sustainable future for humanity.

Conclusions

At the start of this chapter, we referred to the United Nations' Sustainable Development Goals and noted that a majority of the goals are strongly related to food production and security. We argued that to achieve many of these targets represents a huge challenge and will involve concerted multinational action and resolute political will, in both developed and developing countries. However, significant progress has been made in recent decades in improving food security in many areas of the world. The problem now is how to render food production, preparation, consumption, distribution, disposal and replenishment systems sustainable for future generations. The crises that we are currently facing, in terms of differential access to food and the absence of food justice in many countries, issues surrounding excess production and food waste in the industrialised world, and the environmental effects of how we produce and consume food, are so profound that they have the potential to undermine the achievements of the last century entirely. To move towards a sustainable future in the twenty-first century, fundamental changes are urgently required.

Fostering a synergy between top-down, publicly funded interventions and bottom-up local initiatives, which stem from community and individual action and agency, some of which we have referred to in this chapter, is obviously critical. More than this, it is likely that a profound alteration in global food justice to address the inequalities in food entitlement and access, both within and between countries and between the industrialised and developing world, will be required. Whether there is the political will on the part of individual governments and governments working together in international fora, or the necessary popular support for the actions that are needed to achieve this in the coming decades remains to be seen. However, without it, the prospects for achieving food sustainability for humanity appear very bleak indeed.

Iain Young
Institute of Clinical Sciences, University of Liverpool, UK

Paul Collinson
Oxford Brookes University, Oxford, UK

Lucy Antal
The Food Domain, Liverpool

Helen Macbeth
Oxford Brookes University, Oxford, UK

Notes

1. The report has come in for criticism from some quarters for alleged inconsistencies and for recommending a diet which is, in the opinion of some, 'nutritionally deficient' (e.g. Ede 2019; Harcombe 2019).
2. https://www.lovefoodhatewaste.com/ (last accessed 14 February 2019).
3. http://austintexas.gov/environment/zero-waste (last accessed 14 February 2019).
4. https://www.eu-fusions.org (last accessed 14 February 2019).
5. https://www.lastminutemarket.it/english (last accessed 14 February 2019).
6. http://www.fao.org/save-food/news-and-multimedia/news/news-details/en/c/175867/ (last accessed 14 February 2019).
7. https://therealjunkfoodproject.org/ (last accessed 14 February 2019).

References

Abbots, E.-J., and Lavis, A. (eds) (2013) *Why We Eat, How We Eat: Contemporary Encounters between Food and Bodies*, Taylor and Francis, Oxford and New York.

Asgar, M.A., Fazilah, A., Huda,N., Bhat, R., Karim, A.A. (2010) Nonmeat Protein Alternatives as Meat Extenders and Meat Analogs, *Comprehensive Reviews in Food Science and Food Safety*, 9: 513–29.

Baer, H.A., and Singer, M. 2018. *The Anthropology of Climate Change: An Integrated Critical Perspective*, Routledge, Oxford.

Bann, D., Johnson, W., Li, L., Kuh, D., Hardy, R. (2018) Socioeconomic Inequalities in Childhood and Adolescent Body-Mass Index, Weight, and Height from 1953 to 2015: An Analysis of Four Longitudinal, Observational, British Birth Cohort Studies, *The Lancet*, 3(4): 194–203.

Biel, R. (2016) *Sustainable Food Systems: The Role of the City*, UCL Press, London.

Brightman, M., and Lewis, J. (eds) (2017a) *The Anthropology of Sustainability: Palgrave Studies in Anthropology of Sustainability*, Palgrave Macmillan, New York.

Brightman, M., and Lewis,J. (2017b). Introduction: The Anthropology of Sustainability: Beyond Development and Progress. In Brightman, M. and Lewis, J. (eds) *The Anthropology of Sustainability: Palgrave Studies in Anthropology of Sustainability*, Palgrave Macmillan, New York.

Brückner, M., and Caglar, G. (2016) Understanding Meal Cultures – Improving the Consumption of African Indigenous Vegetables: Insights from Sociology and Anthropology of Food, *African Journal of Horticultural Science*, 9: 53–61.

Buckwell, A., and Nadeu, E. (2018) *What is the Safe Operating Space for EU Livestock?* RISE Foundation, Brussels.

Caplan, P. (2016) Big Society or Broken Society? Food Banks in the UK, *Anthropology Today*, 32(1): 5–9.

Carney, M. (2015) *The Unending Hunger: Tracing Women and Food Insecurity across Borders*, University of California Press, Santa Barbara.

Carolan, M. (2015) Re-Wilding Food Systems: Visceralities, Utopias, Pragmatism, and Practice. In Stock, P., Rosin, C. and Carolan, M. *Food Utopias: An Invitation to a Food Dialogue*, London, Routledge, pp.126–139.

Clammer, J. (2017). Cosmpolitanism Beyond Anthropocentrism: The Ecological Self and Transcivilizational Dialogue. Published at: http://sustainability.psu.edu/field-guide/resources/clammer-j-2017-cosmpolitanism-beyond-anthropocentrism-the-ecological-self-and-transcivilizational-dialogue/. Accessed on 12 November 2018.

Collings P., Marten, M.G., Pearce, T., Young, A.G. (2016) Country Food Sharing Networks, Household Structure and Implications for Understanding Food Insecurity in Arctic Canada, *Ecology of Food and Nutrition*, 55(1): 30–49.

Delisle, H., and Batal, M. (2016). The Double Burden of Malnutrition Associated with Poverty, *The Lancet*, 387(10037): 2504–2505.

Devereux, S. (2007) *The New Famines: Why Famines Persist in an Era of Globalization*, Routledge, Abingdon.

EAT-Lancet Commission (2019) *Food in The Anthropocene: the EAT-Lancet Commission on Healthy Diets From Sustainable Food Systems*. Published at: https://www.thelancet.com/commissions/EAT. Accessed on 27 January 2019.

Ede, G. (2019) EAT-Lancet's Plant-Based Planet: 10 Things You Need to Know, *Psychology Today* Blog. Published at: https://www.psychologytoday.com/us/blog/diagnosis-diet/201901/eat-lancets-plant-based-planet-10-things-you-need-know. Accessed on 27 January 2019.

Elton, C. (1927) *Animal Ecology*, Sudgwick & Jackson, London.

Fayaz Bhat, Z., and Fayaz, H. (2011) Prospectus of Cultured Meat: Advancing Meat Alternatives, *Journal of Food Science and Technology*, 48: 125–40.

Feedback (2018) *Feeding Surplus Food to Pigs Safely: A Win–Win for Farmers and the Environment*, Feedback, London.

Fernie, J., Sparks, L. and McKinnon, A.C. (2010) Retail Logistics in the UK: Past, Present and Future, *International Journal of Retail and Distribution Management*, 38(11/12): 894–914.

Food and Agriculture Organization (FAO) (2009) How to Feed the World in 2050. Published at: http://www.fao.org/fileadmin/templates/wsfs/docs/expert_paper/How_to_Feed_the_World_in_2050.pdf. Accessed on 21 September 2018.

Food and Agriculture Organization (FAO) (2017) Food Outlook: Biannual report on Global Food Markets. Published at: http://www.fao.org/3/a-I8080e.pdf. Accessed on 25 September 2018.

Food and Agriculture Organization (FAO) (2018a) No. 3 September: Crop Prospects and Food Situation. Published at: http://www.fao.org/giews/reports/crop-prospects/en/. Accessed on 25 September 2018.

Food and Agriculture Organization (FAO) (2018b) Note on Animal Production. Published at: http://www.fao.org/animal-production/en/. Accessed on 21 September 2018.

Food and Agriculture Organization (FAO), IFAD, UNICEF, WFP and WHO (2017) The State of Food Security and Nutrition in the World 2017. Published at: http://www.fao.org/3/a-I7695e.pdf. Accessed on 12 April 2018.

Godfray, H.C.J., Beddington, J.R., Crute, I.R., Haddad, L., Lawrence, D., Muir, J.F., Pretty, J., Robinson, S., Thomas, S.M. and Toulmin, C. (2010) Food Security: The Challenge of Feeding 9 Billion People, *Science*, 327(5967): 812–18.

Goody, J. (1982) *Cooking, Cuisine and Class: A Study in Comparative Sociology*, Cambridge University Press, Cambridge.

Green, A. and Asinjo, R. (2015) Applying Anthropology to the Campus Dining System: Reflections on Working with Community Food Assessments and the Real Food Challenge, *Practising Anthropology*, 37(2): 22–26.

Gupta, J. and Pouw, N. (2017) Towards a Transdisciplinary Conceptualization of Inclusive Development, *Current Opinion in Environmental Sustainability*, 24: 96–103.

Guyot, J. (2011) Anthropology as Key to Sustainability Science: An Interview with Charles Redman, *Anthropology News*, 52(4): 12.

Hadley, C., and Wutich, A. (2009) Experience-based Measures of Food and Water Security: Biocultural Approaches to Grounded Measures of Insecurity, *Human Organization*, 6(4): 451–60.

Harcombe, Z. (2019) The EAT Lancet Diet is Nutritionally Deficient. Published at: http://www.zoeharcombe.com/2019/01/the-eat-lancet-diet-is-nutritionally-deficient/. Accessed on: 27 January 2019.

Henchion, M., De Backer, C. and Hudders, L. (2017) Ethical and Sustainable Aspects of Meat Production: Consumer Perceptions and System Credibility. In Purslow, P. (ed.) *Meat Quality Aspects: From Genes to Ethics*, Elsevier, Cambridge, MA.

Henchion, M., McCarthy, M., Resconi, V.C., Declan, T. (2014) Meat Consumption: Trends and Quality Matters, *Meat Science*, 98(3): 561–68.

Herrero, M., Thornton, P.K., Notenbaert, A.M., Wood, S., Msangi, S., Freeman, H.A., Bossio, D., Dixon, J., Peters, M., van de Steeg, J., Lynam, J., Parthasarathy, R.P., Macmillan, S., Gerard, B., McDermott, J., Seré, C. and Rosegrant, M. (2010) Smart Investments in Sustainable Food Production: Revisiting Mixed Crop–Livestock Systems, *Science*, 12: 822–25.

Homewood, K.M. (2017) 'They Call It Shangri-La': Sustainable Conservation, or African Enclosures? In Brightman, M. and Lewis, J. (eds) *The Anthropology of Sustainability: Palgrave Studies in Anthropology of Sustainability*, Palgrave Macmillan, New York.

Huis, A. van. (2013) Potential of Insects as Food and Feed in Assuring Food Security, *Annual Review of Entomology*, 58(1): 563–83.

Intergovernmental Panel on Climate Change (2014) Climate Change 2014: Impacts, Adaptation and Vulnerability. Published at: http://www.ipcc.ch/report/ar5/wg2/. Accessed on 21 April 2018.

Jones, D.L., Cross, P., Withers, P.J.A., DeLuca, T.H., Robinson, D.A., Quilliam, R.S., Harris, I.M., Chadwick, D.R. and Edwards-Jones, G. (2013) REVIEW: Nutrient Stripping: The Global Disparity between Food Security and Soil Nutrient Stocks, *Journal of Applied Ecology*, 50: 851–62.

Jörissen, J., Priefer, C. and Bräutigam, K.-R. (2015) Food Waste Generation at Household Level: Results of a Survey among Employees of Two European Research Centers in Italy and Germany, *Sustainability*, 7: 2695–2715.

Latour, B. (2017) Anthropology at the Time of the Anthropocene: A Personal View of What Is to Be Studied. In Brightman, M. and Lewis. J. (eds) *The Anthropology of Sustainability: Palgrave Studies in Anthropology of Sustainability*, Palgrave Macmillan, New York.

Lawrence, G., Lyons, K. and Wallington, T. (eds.) (2010). *Food Security, Nutrition and Sustainability*, London, Routledge.

Leatherman, T., and Goodman, A. (1997) Expanding the Biocultural Synthesis Toward a Biology of Poverty, *American Journal of Physical Anthropology*, 102: 1–3.

Lockyer, J., and Veteto, J.R. (eds) (2013) *Environmental Anthropology Engaging Ecotopia: Bioregionalism, Permaculture and Ecovillages*, Berghahn Books, Oxford.

Melbye, E.L., Onozaka, Y. and Hansen, H. (2017) Throwing It All Away: Exploring Affluent Consumers' Attitudes Toward Wasting Edible Food, *Journal of Food Products Marketing*, 23(4): 416–29.

Milan Urban Food Policy Pact (2015) Text. Published at: http://www.milanurbanfood-policypact.org/text/. Accessed on 27 August 2018.

Milan Urban Food Policy Pact (2017) Food Smart Cities for Development: Recommendations and Good Practices. Published at: http://www.milanurbanfood-policypact.org/2017/02/22/fsc4d-recommendations. Accessed on 1 October 2018.

Murphy, F. and McDonagh, P. (eds) (2016) *Envisioning Sustainabilities: Towards an Anthropology of Sustainability*, Cambridge, Cambridge Scholars Publishing.

Nelson, G.C., Valin, H., Sands, R.D., Havlík, P., Ahammad, H., Deryng, D., Elliott, J., Fujimori, S., Hasegawa, T., Heyhoe, E., Kyle, P., Von Lampe, M., Lotze-Campen, H., Mason d'Croz, D., van Meijl, H., van der Mensrugghe, D., Müller, C., Popp, A., Robertson, R., Robinson, S., Schmid, E., Schmitz, C., Tabeau, A. and Willenbockel, D.. (2014) Climate Change Effects on Agriculture: Economic Responses to Biophysical Shocks, *Proceedings of the National Academy of Sciences of the United States of America*, 111(9): 3274–79.

Nguyen, T.L.T., Hermansen, J.E. and Mogensen, L. (2012) Environmental Costs of Meat Production: The Case of Typical EU Pork Production, *Journal of Cleaner Production*, 28: 168–76.

Nielsen, K.B. (2014) Between Peasant Utopia and Neoliberal Dreams: Industrialisation and Its Discontents in Emerging India. In Wethal, U. and Hansen, A. (eds) *Emerging Economies and Challenges to Sustainability: Theories, Strategies, Local Realities*, Routledge, London.

OLIO (2017) The Problem of Food Waste. Published at: https://olioex.com/food-waste/the-problem-of-food-waste/. Accessed on 13 April 2018.

Peacock, C., and Sherman, D.M. (2010) Sustainable Goat Production: Some Global Perspectives, *Small Ruminant Research*, 89(2–3): 70–80.

Piesse, J., and Thirtle, C. (2009) Three Bubbles and a Panic: An Explanatory Review of Recent Food Commodity Price Events, *Food Policy*, 34(2): 119–29.

Pimentel, D., and Pimentel, M. (2003). Sustainability of Meat-based and Plant-based Diets and the Environment, *American Journal of Clinical Nutrition*, 78(Suppl. 3): 660S–663S.

Piperata, B.A. (2008) Forty Days and Forty Nights: A Biocultural Perspective on Postpartum Practices in the Amazon, *Social Science and Medicine*, 67:1094-1103.

Poore, J., and Nemecek, T. (2018) Reducing Food's Environmental Impacts through Producers and Consumers, *Science*, 260(6392): 987–92.

Quested, T.E., Marsh, E., Stunell, D., and Parry, A.D. (2013) Spaghetti Soup: The Complex World of Food Waste Behaviours, *Resources, Conservation and Recycling*, 79: 43–51.

Reuter, T. (2015) In Response to a Global Environmental Crisis: How Anthropologists are Contributing to Sustainability and Conservation. In Reuter, T. (ed.) *Averting a Global Environmental Collapse: The Role of Anthropology and Local Knowledge*, Cambridge Scholars Publishing, Newcastle upon Tyne.

Schmidhuber, J., and Tubiello, F.N. (2007) Global Food Security under Climate Change, *Proceedings of the National Academy of Sciences*, 104(50): 19703–19708.

Secondi, L., Principato, L., and Laureti, T. (2015) Household Food Waste Behaviour in EU-27 Countries: A Multilevel Analysis, *Food Policy*, 56: 25–40.

Sen, A. (1981) *Poverty and Famines: An Essay on Entitlement and Deprivation*, Oxford University Press, Oxford.

Stevenson, E., Green, L.E., Maes, K.C., Ambelu, A., Tesfaye, Y.A, Rheingans, R. and Hadley, C. (2012) Water Insecurity in 3 Dimensions: An Anthropological Perspective on Water and Women's Psychosocial Distress in Ethiopia, *Social Science and Medicine*, 75: 392–400.

Tan, Z.X., Lal, R. and Wiebe, K.D. (2005) Global Soil Nutrient Depletion and Yield Reduction, *Journal of Sustainable Agriculture*, 26(1): 123–46.

Tanumihardjo, S.A., Anderson, C., Kaufer-Horwitz, M., Bode, L., Emenaker, J., Haqq, A.M., Satia, J.A., Silver, H.J. and Stadler, D.D. (2007) Poverty, Obesity, and Malnutrition: An International Perspective Recognizing the Paradox, *Journal of the American Dietetic Association*, 107(11): 1966–72.

Taylor, K. (2017) These 10 Companies Control Everything You Buy, *Business Insider UK*. Published at: http://uk.businessinsider.com/10-companies-control-food-industry-2017-3/#kelloggs-1. Accessed on 20 January 2019.

Thornton, P.K. (2010) Livestock Production: Recent Trends, Future Prospects. *Philosophical Transactions of the Royal Society London B, Biological Sciences,* 365(1554): 2853–2867.

United Nations Department of Economic and Social Affairs (2015) 17 Sustainable Goals: 17 Partnerships. Published at: https://sustainabledevelopment.un.org/content/documents/211617%20Goals%2017%20Partnerships.pdf. Accessed on 12 October 2018.

United Nations (2018) About Sustainable Development Goals. Published at: https://www.un.org/sustainabledevelopment/sustainable-development-goals/. Accessed on 17 December 2018.

Waal, A. de (1997) *Famine Crimes: Politics and the Disaster Relief Industry in Africa*, James Currey, Oxford.

Waal, A. de (2005) *Famine that Kills: Darfur, Sudan*, Oxford University Press, Oxford.

Walsh, B. (2013) The Triple Whopper Environmental Impact of Global Meat Production. *Time*, 16 December 2013. Published at: http://science.time.com/2013/12/16/the-triple-whopper-environmental-impact-of-global-meat-production/. Accessed on 21 April 2018.

Wheeler, T., and von Braun, J. (2013). Climate Change Impacts on Global Food Security, *Science*, 341(6145): 508–13.

WRAP (2018) Food Futures: From Business as Usual to Business Unusual. Published at: http://www.wrap.org.uk/content/food-futures. Accessed on 19 September 2018.

Wutich, A., and Brewis, A. (2014) Food, Water and Scarcity: Toward a Broader Anthropology of Resource Insecurity, *Current Anthropology*, 55(4): 444–68.

Ziv, G., Baran, E., Nam, S., Rodríguez-Iturbe, I., Levin, S.A. (2012) Trading-off Fish Biodiversity, Food Security, and Hydropower in the Mekong River Basin, *Proceedings of the National Academy of Sciences* (March).

CHAPTER 2
FOOD INSECURITY AND SUSTAINABILITY IN SUB-SAHARAN AFRICA

Paul Collinson

Introduction

The aims of this chapter are to provide an overview of the extent of current food insecurity in sub-Saharan Africa, to explore the interaction between different drivers of food insecurity in this region and to underline the centrality of conflict and displacement in driving food insecurity and affecting food sustainability in a number of African countries. The chapter is divided into four sections. After a brief outline of food sustainability, the first section provides an overview of undernourishment trends in sub-Saharan Africa. The second section examines the importance of different drivers in creating hunger in the region, and the implications of this for food sustainability. The chapter then goes on to discuss the specific example of farmer–herder conflicts, presenting an overview of some of the research that has been carried out in this area. The discussion ends with the conclusion that resolving conflict is the key to creating sustainable food systems in many sub-Saharan African countries, based on an understanding of the interrelated environmental, political, economic and social dynamics which drive conflict in this region.

What Is Food Sustainability?

The subject of this book is food *and* sustainability. This chapter has a somewhat narrower focus, however: 'food sustainability'. As the definition provided in the Introduction to this volume implies, food sustainability encompasses a number of interrelated themes and aims. Self-evidently, it implies food security – the ability to produce or procure enough food to meet a population's nutritional requirements – which is sustained over time. The environmental dimension is critical: systems of production and grazing should ideally have a neutral impact on the environment, to preserve it for future generations. It also involves equitable access to food, rather than one group being favoured over another. The emphasis here is upon issues of social justice surrounding food entitlement rather than volume of food produced per se; it is thus invariably anti-Malthusian in approach. Finally, sustainability usually has a political aspect in the sense that implementing sustainable solutions implies some form of political change.

On the latter point, it seems to be widely accepted that democratic systems of governance are more likely to be associated with policies promoting sustainability than autocracies. Likewise, maximal political devolution is more likely to promote 'bottom-up', participatory and sustainable methods of organising food production than 'top-down', externally or centrally imposed policies. However, there are exceptions here in the sense that governments often need to come together to promote collective policies aimed at environmental protection over shared areas and to avoid a 'tragedy of the commons' (Hardin 1968) situation emerging; the environmental measures of the EU are perhaps the most obvious example in this regard.

For the most part, the environmental impact of community-organised farming or grazing systems is likely to be less deleterious than large-scale cash cropping or intensive livestock raising controlled by multinational corporations (Biel 2016). This may not always hold true, however, since overgrazing by pastoralists in sub-Saharan Africa has become a serious problem in recent years – a subject that I will return to later in the chapter.

Undernourishment Trends

Nowhere is the quest for food sustainability more pressing than in sub-Saharan Africa. Although Asia had the highest number of hungry people in the world in 2016 – 520 million, compared to 243 million in Africa (Food and Agriculture Organization et al. 2017: 7) – the prevalence of undernutrition[1] is higher in sub-Saharan Africa than in any other region; at 22.7 percent of the population it is almost twice the level of that found in Asia, and more than double the global average. Despite some notable successes over the past two to three decades, food insecurity in many sub-Saharan African countries is proving stubbornly persistent, and has increased in several in recent years (see Figure 2.1 and Tables 2.1 and 2.2). Sub-Saharan African countries are consistently towards the bottom of the Global Hunger Index (von Grebmer et al. 2017), with the Central African Republic and Chad occupying the lowest two places in 2017.[2] The number of people displaced by conflict worldwide is also overwhelmingly concentrated in sub-Saharan African countries (Internal Displacement Monitoring Centre 2016).

Analysis of trends in undernourishment in different sub-regions of sub-Saharan Africa is revealing (Figure 2.1). From 1990 to 2016, the proportion of the population suffering from undernourishment fell by 10 percentage points across the region as a whole, and decreased in every sub-region apart from in

Table 2.1 Prevalence of undernutrition worldwide, 2016. Adapted from table in Food and Agriculture Organization et al. 2017, p. 6.

Region	Prevalence of Undernutrition 2016, percent
World	11.0
Sub-Saharan Africa	22.7
Asia	11.7
North Africa	8.3
Oceania	6.8
Latin America and Caribbean	6.6
North America and Europe	< 2.5

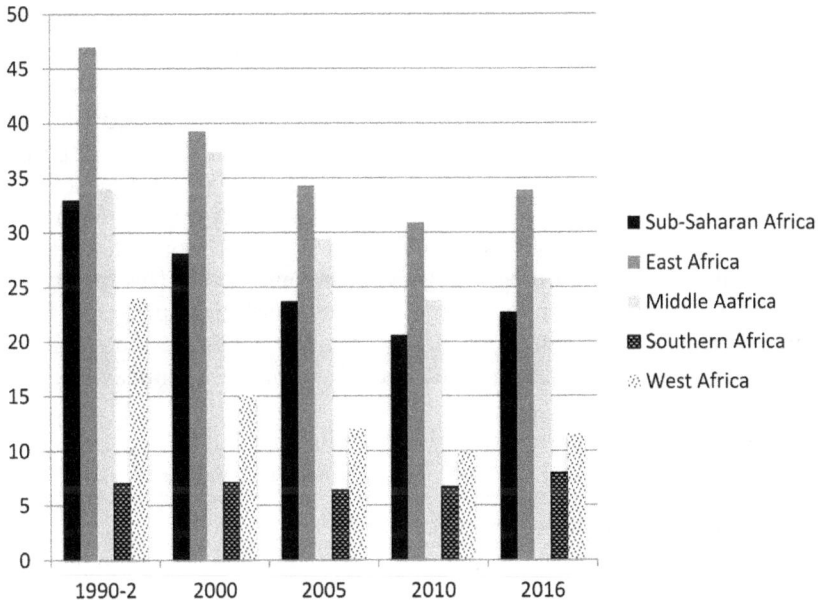

Figure 2.1 Undernourishment trends, 1990–2016, percentage of population. Figure created by the author.

Sources: Data compiled from Food and Agriculture Organization 2015 and Food and Agriculture Organization 2017b.

Southern Africa, where it increased from 7 to 8 percent. The biggest fall was in East Africa, 13.1 percent, with a 12.5 percent fall in West Africa. In Middle Africa,[3] a decrease of 8.2 percent was recorded. However, it can be seen that there has been an increase in undernutrition from 2010 to 2016 in every sub-region of between 1.3 (Southern Africa) and 3 (East Africa) percentage points, reversing the previous trend.

A combination of factors is at play here. The Food and Agriculture Organization of the UN attributes the renewed upward curve to adverse climatic conditions (e.g. El Niño weather patterns, drought), conflict and adverse global economic conditions (Food and Agriculture Organization 2015; Food and Agriculture Organization 2017b: 1). It is also notable that the two sub-Saharan African countries whose populations are suffering from the highest levels of food insecurity at the present time, South Sudan and the Central African Republic – 48 percent and 36 percent respectively (see Table 2.2) – have also been the locations for the most intense internal conflicts over the past five years. The figures for individual countries also mask significant sub-national variations, with areas such as northern Nigeria, the Lake Chad basin, northern Mali and western and southern Sudan suffering far higher levels of food insecurity than their countries as a whole. These points are discussed in more detail below.

Drivers of Food Insecurity

Several interrelated factors can be identified as being associated with food insecurity in sub-Saharan Africa (Table 2.3). Although they are separated out neatly here, the situation in reality is far messier. Virtually all of these drivers are intimately interconnected, inhibiting or reinforcing each other and criss-crossed by complex multipliers and feedback loops, and changing dynamically and often unpredictably over time and space. This observation alone demonstrates the importance of adopting a holistic approach to implementing measures to address food sustainability, whilst at the same time underlining the huge challenges which are inherent in this endeavour.

Lack of space precludes a comprehensive discussion of all of these drivers, but some of them are addressed in more detail below.

Economic and Political Drivers

Poverty

The countries of sub-Saharan Africa are some of the poorest in the world, as measured by GDP per capita and through the Human Development Index (Table 2.4).

Table 2.2 Food insecurity in selected countries in sub-Saharan Africa.

Country	Food insecurity (IPC Phase 2–5, unless otherwise stated)[a]	Percentage of population[b]	Position on Global Hunger Index 2017 (out of 119 countries)[c]
Burkina Faso	620,000[d]	3.0	92
Cameroon	3.9m[e]	10.4	74
Central African Republic	2.0m[f]	35.6	119
Chad	3.5m[g]	33.1	118
Mali	4.1m[h]	22.9	94
Mauritania	602,000[i]	16	83
Niger	1.4m[j]	7.2	111
Nigeria	5.3m[k]	1.9	84
Senegal	314,000[l]	2.1	67
Somalia	6.2m[m]	50.4	ND[p]
South Sudan	6.3m[n]	48.3	ND[q]
Sudan	4.8m[o]	13.0	113

[a] Integrated Food Security Phase Classification (IPC) Phase 2 ('Stressed'); Phase 3 ('Crisis'); Phase 4 ('Emergency'); Phase 5 ('Catastrophe' / 'Famine'). The IPC is the generally accepted standard measure for food insecurity worldwide. It was developed by the UN Food and Agriculture Organization in 2004 (IPC 2018).
[b] All population figures from Central Intelligence Agency World Factbook 2018.
[c] Von Grebmer et al. 2017.
[d] UNOCHA 2018a.
[e] USAID 2018.
[f] Data from USAID 2017, as of September 2017.
[g] UNOCHA 2018b: 10.
[h] UNOCHA 2018d; UNHCR 2018.
[i] UNOCHA 2018a.
[j] Ibid.
[k] Cadre Harmonisé 2018 [Figure given is IPC Phases 3–5 for 16 states].
[l] As of November 2017; World Food Programme 2018a.
[m] World Food Programme 2018b.
[n] World Food Programme 2018c.
[o] UNOCHA 2018c.
[p] Not ranked due to lack of data.
[q] Not ranked due to lack of data.

Table 2.3 Principal drivers of food insecurity in sub-Saharan Africa.

Political	Economic	Social / Demographic	Environmental
• Weak governance	• Poverty	• Population growth	• Climate change
• Political instability	• Declining food production	• Conflict	• Desertification
• Lack of democracy	• Economic instability	• Changing migration patterns	• Land stress
• Marginalisation / disenfranchisement of groups / communities	• Food price rises	• Displacement	• Soil erosion / degradation
• Corruption	• Public debt	• Ethnic / tribal tensions	• Water shortages
• Weakness of civil society	• Cash cropping	• Youth bulge	• Pests; livestock, crop diseases
• Weak / corrupt justice systems (implications for land ownership rights)	• Security sector spending crowding out other sectors	• Gender inequality	• Ecosystem changes
	• Weak financial governance	• Social marginalisation of specific groups / communities	
		• Low human development indicators	

Poverty is a significant driver of food insecurity. It is a primary determinant of the ability of people to afford an adequate supply of food, and also restricts the opportunities available for people to increase their status and life chances (Luan et al. 2013). Poverty depresses life expectancy and renders people more susceptible to illness, which also impacts on food security (Maunder and Wiggins 2007). In sub-Saharan Africa, rural poverty has been a major driver of rural–urban migration in recent decades (Barrios et al. 2006), a trend strongly associated with a movement away from subsistence farming and towards pre-prepared food consumption (Bourne et al. 2002; Codjoe et al. 2016). This has clear implications for food sustainability.

Table 2.4 GDP per capita ranking (2017, est.) and Human Development Index 2015, selected countries (Sources: Central Intelligence Agency World Factbook 2018; UNDP 2017). In a number of these countries, GDP per capita figures are inflated by oil revenues; examples include Chad, Sudan and South Sudan. The Human Development Index is, therefore, a more accurate reflection of poverty.

Country	GDP per capita rank 2017 position (out of 228 countries)	Human Development Index[a] 2015 position (out of 188 countries)
Central African Republic	228	188
Burundi	227	184
DRC	226	176
Niger	222	187
Eritrea	220	179
South Sudan	219	181
Burkina Faso	208	185
Mali	202	175
Chad	198	186
Senegal	196	162
Cameroon	187	153
Mauritania	174	157
Sudan	173	165
Nigeria	163	152

[a] The Human Development Index is a composite figure calculated from life expectancy, education and standard of living indicators.

Food Production

Food sustainability ideally includes a dimension of self-sufficiency. Very few countries in sub-Saharan Africa are self-sufficient in food. The total volume of agricultural production (meat and primary crops) across the region has risen at less than 1 percent per year since the 1960s, with even smaller rises recorded in fish production, far below the increase in demand for food over this period (Cornia et al. 1992: 195; Chauvin et al. 2012: 4–6). The region became a net importer of agricultural products in the early 1980s (Rakotoarisoa et al. 2011: 5) and since then has witnessed an exponential growth in the amount of food imports (Djurfeldt et al. 2005: 2). In 2017, sub-Saharan Africa imported 20 percent of the needs of its population for major food commodities (OECD–Food and Agriculture Organization 2018: 46), despite having around a quarter of the world's total arable land (Bafana 2014: 9). The gap between food production and food imports is likely to increase further in the coming decades, with an expected tripling of demand for cereals by 2050 (van Ittersum et al. 2016).

Importing more food than is produced renders populations more vulnerable to global price shocks, which can also cause instability and conflict (Arezki and Brückner 2011). Evidence suggests that the food price spikes that occurred between 2008 and 2011 were strongly correlated to food riots and social unrest in a number of countries (e.g. Collinson and Macbeth 2014: 5; Ecker 2014; Biel 2016: 14; Food and Agriculture Organization et al. 2017: 52). High levels of food imports also contribute to levels of dependency upon international aid. Over time, people lose the knowledge and skills necessary to grow their own food and become tied into the global humanitarian system.

Weak / Poor Governance

Weak and poor governance is a feature of many sub-Saharan African countries. A lack of democracy, political oppression and instability, endemic corruption, weak penetration of governmental authority, the prevalence of local warlords and underdeveloped systems of local government, among several other factors, all contribute to food insecurity, to varying degrees.

As Sen (1981) demonstrated, the critical factor in creating food security is not availability of food per se but *entitlement* to food. Governments which prioritise the needs of certain groups over others, that are undemocratic or autocratic (or, as with some African countries, kleptocratic) and/or are in conflict with sectors of their own populations – all too often the case in this region – are highly unlikely to pursue policies which promote equitable access to food across their territories. On the contrary, a number of regimes in the region have actively used access to food as a weapon of war, starving communities opposed to them in order to submit them to their authority as a crude and brutal counter-insurgency strategy.

Corruption is a particular issue in sub-Saharan Africa. In Transparency International's latest index of perceptions of corruption in 180 countries, 17 of the 30 countries perceived to be the most corrupt are located in the region, including three of the bottom five – Sudan, South Sudan and Somalia (Transparency International 2018). Corruption has a significant impact on food security and sustainability by making it more difficult and more costly for international organisations, including foreign companies, to operate in the country, as well as contributing to a leaching of humanitarian and development aid. Although most donors bypass governments by ensuring that populations in need are targeted as directly as possible, official corruption inevitably contributes to the redirection of some of this funding away from those who need it most.

Social / Demographic Drivers

Population Growth

Whilst crude Malthusian arguments can be rejected as the *sole* explanation of food insecurity, it is nonetheless necessary to acknowledge population growth as a significant contributory factor – not least because of its impact on a number of other drivers, such as soil degradation and competition over land, and also because of the scale of population growth anticipated in sub-Saharan Africa over the coming decades.

Population growth in sub-Saharan Africa is higher than in any other region of the world at the present time, and its population is expected to double by 2050 to almost two billion (*Guardian* 2013; United Nations Department of Economic and Social Affairs 2017). At current rates of food production, it has been estimated that only 13 percent of the region's food requirements will be met by 2050 (Global Harvest Initiative 2012: 15; Munang and Andrews 2014: 6). This is likely to exacerbate food insecurity, mass migration, environmental degradation and conflict.

In the face of such dire predictions, there have been calls for a 'sustainable intensification' of agricultural production: a 'Green Revolution for Africa' (Toenniessen et al. 2008; Conceição et al. 2016). It is to be hoped that this can be achieved by avoiding some of the deleterious effects which befell the Green Revolution in Asia (c.f. Djurfeldt et al. 2005, 2006). It will, therefore, need to be accompanied by measures to ensure that it does not occur at the expense of environmental stress, such as loss of soil nutrients, depletion of water sources, emergence of new diseases and pests and a shrinkage in bio-diversity (Reynolds et al. 2015) nor lead to small farmers being drawn en masse into global food chains which increase dependency on global markets and vulnerability to price shocks. The problems of large-scale cash cropping are well documented and may be viewed as the antithesis of what is generally associated with food sustainability. The role of Western agricultural corpo-rations in supplying agricultural inputs (especially seeds) to farmers in the developing world has also attracted much controversy in allegedly driving increasing levels of debt. As Biel (2016: 116–17) has pointed out, the Green Revolution can potentially be blamed for a significant proportion of the high rates of suicide among Indian farmers.

The Green Revolution measures that have already been introduced in some countries across sub-Saharan Africa have had mixed results.[4] Agricultural intensification is also unlikely to work in isolation, and will probably need to be accompanied by other strategies[5] aimed at poverty reduction and increasing non-agricultural employment opportunities (Frelat et al. 2016) in order to be in any way sustainable.

Insecurity and Conflict

Many countries in sub-Saharan Africa, particularly in the Sahel[6] and Central Africa, are suffering from internal conflict and insecurity. As previously mentioned, the most serious conflicts in the region in the past decade have been in South Sudan and the Central African Republic, where civil wars have had a devastating impact on the humanitarian situation in both countries and have been almost wholly responsible for the huge increases in food insecurity that have been recorded over this period (Food and Agriculture Organization et al. 2017). A famine was declared by the UN in parts of South Sudan for several months in 2017, the only place in the world where a famine has emerged since the Somalian food crisis of 2011. Insecurity is also prevalent in the Democratic Republic of Congo (DRC), in parts of Somalia, in northern Nigeria, in northern Mali, in parts of Sudan, in western Niger and in the Lake Chad basin, among other locations, with a concomitant impact on food security and sustainability in these areas.

The impact of conflict on food security and nutrition in sub-Saharan Africa has been very widely documented (for specific country examples, see Bundervoet et al. (2009) in relation to Burundi; Akresh et al. (2010) in relation to Rwanda; Minoiu and Shemyakina (2012) in relation to Cote d'Ivoire; Kinyoki et al. (2017) in relation to Somalia, among countless other studies). The causal relationship between the two operates in several interrelated ways (c.f. Brinkman and Hendrix 2011), some of which I consider below.

Firstly, conflict has caused unprecedented levels of population displacement across the region. Many countries are hosting refugees from neighbouring countries, which in some cases has led to tensions with host populations and put a strain on local economies. There are also significant levels of internal displacement. Secondly, conflict draws (primarily young) men into armed forces or militia groups, taking them away from their families and preventing them from participating in agricultural activities. The localised nature of conflict in sub-Saharan Africa means that high casualty rates among young men can also have devastating effects on specific communities. Thirdly, conflict degrades or destroys agricultural assets and infrastructure (Deininger and Castagnini 2006), and also has environmental effects through the legacy of unexploded ordnance, which can take significant areas of otherwise productive land out of agricultural use (Oyeniyi and Akinyoade 2014). Fourthly, conflict often leads to combatants turning to local agricultural resources for their own sustenance, which places greater pressure on community food resources as well as increasing the level of attacks against civilians (Koren and Bagozzi 2017). Fifthly, conflict in the region is also usually accompanied by sexual violence, primarily against women and girls; this compounds feelings of insecurity among civilian populations and has a suppressive effect on female participation rates in agricultural work. Sixthly, conflict affects the ability of populations to access local markets due to insecurity and/or degradation of the physical infrastructure, which also has the effect of increasing local food prices (Breisinger et al. 2014: 2). Finally, conflict serves to direct the resources of governments towards their armed forces and the procurement of military equipment.

On the last point, the disproportionate share of government budgets that are devoted to the security and military sectors across the region is both a symptom and a cause of conflict. This inevitably crowds out spending on other areas of government activity, such as agricultural investment, which (potentially) influences food security and renders the populations more dependent upon humanitarian aid provided by international public agencies such as the UN and non-governmental organisations, with governments themselves contributing little to this support.

It is also clear that food insecurity itself causes, exacerbates or prolongs conflict, thereby creating a multiplier effect and complicating attempts at resolution (Pinstrup-Andersen and Shimokawa 2008).

Displacement and Migration

Displacement rates, caused largely by conflict and exacerbated by environmental factors (see below), have risen markedly in several countries in this region over recent years. Clearly displacement is a very significant driver of food insecurity: put simply, if people are forced to leave their home areas, they cannot grow their own food. Displacement also increases dependency by concentrating people in camps for Internally Displaced Persons (IDPs) and for refugees, where they become reliant on food aid for survival (see also Kaiser this volume). The longer that people spend in such camps, the less likely it is that they will return to their home areas. Traditional knowledge of food production, cultivation and preparation techniques declines and eventually dissipates altogether, thus increasing dependency. Once established, this vicious cycle becomes difficult to break – even in areas where security has improved, long-term residents of displacement camps remain reluctant to return to their homes. Although huge efforts are being made by international agencies in many countries in sub-Saharan Africa to move IDPs and refugees back to their places of origin, progress is slow and in many places is stymied by conflict, drought and environmental change, which is precipitating even more displacement.

Levels of population displacement across the Sahel and Central Africa have increased significantly in recent years. In the Sahel[7], for example, the number of displaced people has almost tripled in just four years, from 1.4 million in

2014 to 5 million in 2018, precipitated largely by conflict (UNOCHA 2018a: 7; Internal Organization for Migration 2018). As of 2015, 80 percent of the world's refugee population originated from just five countries: Central African Republic, Democratic Republic of the Congo, Somalia, South Sudan and Sudan (Kasozi 2017). Significant proportions of the populations in the Central African Republic and South Sudan are displaced: 20 percent (USAID 2017) and 30 percent respectively (UNOCHA 2018e). Sometimes, displacement can be extremely rapid. For example, the world's largest refugee camp is currently the Bidi Bidi site in northern Uganda, which was established in August 2016 for refugees fleeing the civil war in South Sudan. Just a few months later, in early 2017, its population had reached 270,000 (*Guardian* 2017).

Farmer–Herder Conflicts

In many countries in sub-Saharan Africa, social tensions and resource competition between settled and nomadic communities have been an important driver of conflict for several decades – and probably long before that (see, for example, Blench 1996; Breusers et al. 1998; Tonah 2006; Olaniyan 2015). Indeed, it could be argued that this is the most important single driver of conflict in this region. However, it would be incorrect to argue that 'livelihood conflicts' are somehow an intrinsic and endemic aspect of sub-Saharan African societies or that there is something inevitable and 'natural' about them – not least because of strong evidence that they are increasing in number and intensity (e.g. Nori et al. 2005; Moritz 2010; De Haan et al. 2016; Cabot 2017; *The Economist* 2017).

Extensive research on farmer–herder conflicts emphasise their multidimensional aspects, and the fact that each conflict is highly specific to its own social, cultural, political and geographical context. Several, often interrelated, drivers are usually at play. For example, government policies are sometimes blamed, in that public authorities in several countries across the region are recognised as favouring settled communities and neglecting the needs of pastoralist communities or instrumentally marginalising them (Schilling et al. 2010; Benjaminsen et al. 2012). But climate-induced or aggravated environmental change is undoubtedly a significant factor in exacerbating tensions, competition and conflict between different communities (e.g. Shettima and Tar 2008; Fjelde and Uexkull 2012; Brottem 2016) – partly because it leads to overgrazing by, and/or the geographical expansion of, pastoralist groups (see below). Conflict also disrupts pastoralists' migration routes and traditional modes of production. This may lead them to be drawn into seeking alternative livelihoods, such as involvement in smuggling and criminality, particularly in northern areas of the Sahel (De Haan et al. 2016).

Another factor that is relevant here is the terms of trade between pastoralists and farming communities. Nori et al. (2005) make the important point that in times of plenty, market exchanges generally favour pastoralists, at least theoretically, since the calorific value of the food they buy (largely cereals) tends to be greater than the food they sell (largely dairy products). As an example, camel milk has around 700 kilocalories per kilo, whereas rice and wheat contain around 3,000–3,500 kcal/kg (ibid. :12). However, in times of stress caused by drought, disease, conflict or market failure, terms of trade are often reversed. Nori et al. provide the example of northern Somalia in 2000, when drought conditions and a cattle-import ban imposed by Arab countries caused the purchasing power of pastoralists to fall by two-thirds in one year (ibid.: 12–13).

Some of the better-known conflicts that have occurred on the continent in recent decades have involved farming and herding communities, although they have not necessarily been recognised as such. Thus, the Rwandan genocide of 1994, although primarily an ethnic conflict, had its origin partly in the differing livelihoods of Hutu and Tutsi communities. Not just ethnic labels, the terms translate as 'people who farm' and 'people who own cattle' respectively (Ejigu 2009: 891). A number of commentators have pointed out that food scarcity and resource competition between the two groups was an important dimension of one of the most shocking and tragic episodes in modern African history (e.g. Ohlsson 1999, cited in Cabot 2017). Similarly, Mazo (2010) asserts that the ongoing conflict in the Darfur region of Sudan is also driven by tensions between settled and nomadic communities, rather than being simply an 'Arab versus African' conflict, as it is commonly portrayed.

Examples of other current conflicts between farmers and pastoralists across this region include:

- Fulani nomads and settled, particularly Biafran, communities in Nigeria. The expansion south by Fulani groups has brought them into competition with farmers, which led to an estimated 2,500 deaths in 2016 (International Crisis Group 2017).
- Peul pastoralists and farming communities in the Central African Republic. The civil war that erupted in 2013 when the government was overthrown by a coalition of militant groups has an important livelihood dimension. Herders are largely Muslim while farmers are largely Christian, a factor that played a role in driving sectarian violence during the war (Conciliation Resources 2015: 16–17).
- Shilluk (farming) and Dinka (pastoralist) communities in north-eastern South Sudan, with Dinka herders increasingly encroaching onto Shilluk land. This has fuelled the civil war-related violence in the area since 2015, as the government has sided with Dinka pastoralists to push Shilluk militants and civilians out of several settlements on the west bank of the Nile.

It is worth noting that many of these conflicts are deep-seated. Some were documented many years ago by anthropologists working in the region, such as Edward Evans-Pritchard ([1948] 2014) and Godfrey Lienhardt ([1954] 1999). However, conflict did not feature strongly in the ethnographic focus of the colonial era, with kinship, religion and even food being more prominent subjects of study. One of the key differences today in comparison with the past is the availability and sophistication of weaponry, which means that the human effects of their use are far more lethal.

Environmental Drivers

The conflicts described above have undoubtedly been exacerbated by land stress and climate change in recent years. An increasing lack of grazing land for pastoralist communities is a particular problem across the entire region and that, as well as the increasing size of herds, is exacerbating long-term tensions with settled populations as herders encroach upon agricultural land.

Land degradation and desertification are increasing across the region, removing more and more land from food production (United Nations Economic and Social Council 2017). Overgrazing, land-use changes (particularly deforestation), population pressures, agricultural intensification and climate change, due both to anthropogenic factors and as a result of natural variations, are the principal causes (e.g. World Bank 2014). The region is recognised as being particularly vulnerable to the effects of climate change, due to a combination of geographical, environmental, political and socio-economic factors (Connolly-Boutin and Smit 2016), and as part of a continent that is likely to be the most adversely affected by climate change worldwide (Intergovernmental Panel on Climate Change 2007).

Increasing soil degradation and desertification, along with a reduction of water availability and quality, almost certainly climate-change related, have led to a geographical expansion in both cultivated and pasture land over recent years. Nomadic populations are forced to move increasing distances in search of suitable grazing land (e.g. Bauer 2011), while farmers need to cultivate over wider areas as existing plots and fields become unproductive. In West and Central Africa, seasonal transhumance migrations have exhibited an increasing southerly shift over recent years (e.g. Tonah 2006; Ofuoku and Isife 2009; Agyemang 2017). This has brought some communities into conflict (de Bruijn and Van Dijk 2003; *The Economist* 2017). Another trend here is changing land use through publicly funded irrigation schemes, which has also reduced the land area available for pastoralist communities in several countries (Nori et al. 2005: 15).

Conclusion: Towards Sustainability?

Much of the social science research on food sustainability emphasises the political implications of implementing sustainable solutions. A focus on social justice combined with an expansion of democratic representation are seen as critical to improving access to food which is universal and enduring. To this end, a plethora of grass-roots organisations have emerged in recent years in both Western and developing nations under the 'food democracy' umbrella (Norwood 2015).[8] Some examples are provided elsewhere in this volume.

Despite this, the road to food sustainability in both the developed and developing world is likely to prove extremely rocky. In relation to most Western democracies, the introduction of more sustainable food production systems is at least a prominent part of the political discourse, and even if it has not become a central component of the political agenda everywhere, it is possible to envisage significant change occurring over the coming decades. However, in the context of sub-Saharan Africa, this seems an almost impossible pipe dream. Although there are some glimmers of light in the democratic transition in Nigeria in 2016 and relatively peaceful transfers of power in Kenya and Ethiopia in 2018, perhaps the best that can be hoped for at present is that some countries move towards more democratic systems of governance involving free and fair elections and at least a semblance of competent, non-corrupt and professional public services over the coming decades. This *may* lead to more equitable and sustainable systems of food production, distribution and access – and less conflict. However, it is, if anything, more likely that many countries will regress further as the challenges they face in feeding their people, some of which I have outlined here, prove insurmountable (even if the political will is present in the first place) and competition between opposed groups and communities increases. Given this, the urgency to understand and address the problems of undernutrition and food insecurity in poorer nations becomes ever more acute – an endeavour in which detailed micro-studies involving significant inputs from anthropologists and other social scientists will have an important role.

Food sustainability in sub-Saharan Africa is obviously a long way off. The available evidence suggests that a multi-sectoral approach is required addressing most, if not all, of the drivers of hunger. As I have emphasised, insecurity and conflict, particularly between herders and farmers, are absolutely fundamental to the escalating food crises in many areas in this region and have a knock-on effect on virtually all the other factors. This leads to the conclusion that conflict resolution and social reconciliation will be critical to improving food security across much of sub-Saharan Africa. Without this, creating enduring food sustainability in this region going forward will be very difficult, if not impossible, to achieve.

Paul Collinson
Oxford Brookes University, Oxford, UK

Notes

1. Defined as 'the status of persons, whose food intake regularly provides less than their minimum energy requirements' (Food and Agriculture Organization 2017a).
2. The index is somewhat distorted due to a lack of comparable data for several countries, including South Sudan, Syria and Yemen.
3. Middle Africa is defined by the UN as encompassing Angola, Cameroon, the Central African Republic, Chad, the Democratic Republic of the Congo, Equatorial Guinea, Gabon, the Republic of the Congo, and São Tomé and Príncipe.
4. For example, research by Dawson et al. (2016) found that the imposition of 'enforced modernization' on smallholders in western Rwanda only benefited a wealthy elite, and that the policies were actually contributing to landlessness and inequality among the rural poor.
5. One example is 'Conservation Agriculture', which has been promoted as a way to pursue an intensification of production sustainably by focusing on minimising tillage, maintaining soil cover, diversifying crop rotations and using appropriate fertiliser (Vanlauwea et al. 2014). Studies have demonstrated that the introduction of social protection measures can also have a positive impact in improving food security and sustainability in situations of environmental stress. Research by Devereux, for example, has shown that initiatives such as weather-indexed insurance, public works programmes, agricultural input subsidies and buffer stock management can help to iron out the often extreme fluctuations in income and access to food among small farmers between seasons or between 'good' and 'bad' years (Deveraux 2016).
6. The Sahel is generally defined as a semi-arid belt to the immediate south of the Sahara. Its geographic extent is disputed; according to some definitions, it stretches from the Atlantic to the Red Sea, while under other definitions it is essentially coterminous with francophone West Africa and stops at the border of Chad and Sudan. The former, maximal definition is implied here.
7. Defined in UNOCHA 2018a as Burkina Faso, Chad, Mali, Mauritania, Niger, Cameroon, Nigeria (Adamawa, Borno and Yobe states) and Senegal
8. Examples include organisations such as Food Democracy Now!, the Food Sovereignty Alliance and the US Food Councils Network (Carlson and Chappell 2015).

References

Agyemang, E. (2017) Farmer–Herder Conflict in Africa: An Assessment of the Causes and Effects of the Sedentary Farmers–Fulani Herdsmen Conflict. A Case Study of the Agogo Traditional Area, Ashanti Region of Ghana. Master's Thesis, University of Agder. Published at: https://brage.bibsys.no/xmlui/bitstream/handle/11250/2459940/UT-505%20-%20Agyemang%2C%20Emmanuel.pdf?sequence=1. Accessed on 30 March 2018.

Akresh, R., Verwimp, P. and Bundervoet, T. (2010) Civil War, Crop Failure and Child Stunting in Rwanda. *Economic Development and Cultural Change*, 59(4): 777–810.

Arezki, R. and Brückner, M. (2011) Food Prices, Conflict and Democratic Change. International Monetary Fund Working Paper 1162, International Monetary Fund, Washington, D.C.

Bafana, B. (2014) Denting Youth Unemployment Through Agriculture. In Tafirenyika, M. (ed.) *Africa Renewal 2014 Special Edition*, United Nations, New York, pp. 8–9.

Barrios, S., Bertinelli, L. and Strobl, E. (2006) Climate Change and Rural–Urban Migration: The Case of Sub-Saharan Africa, *Journal of Urban Economics*, 60: 357–371.

Bauer, S. (2011) Stormy Weather: International Security in the Shadow of Climate Change. In Brauch, H.G., Oswald Spring, U., Mesjasz, C., Grin, J., Kameri-Mbote, P., Chourou, B., Dunay, P. and Birkmann, J. (eds) *Coping with Global Environmental Change, Disasters and Security: Threats, Challenges, Vulnerabilities and Risks.* Hexagon Series on Human and Environmental Security and Peace, vol. 5. Springer, Berlin–Heidelberg–New York, pp. 719–733.

Benjaminsen, T.A., Alinon, K., Buhaug, H. and Buseth, J.T. (2012) Does Climate Change Drive Land-Use Conflicts in the Sahel? *Journal of Peace Research*, 49(1): 97–111.

Biel, R. (2016) *Sustainable Food Systems: The Role of the City*, UCL Press, London.

Blench, R.M. (1996) Pastoralists and National Borders in Nigeria. In Nugent, P. and Asiwaju, A.I. (eds) *African Boundaries: Barriers, Conduits and Opportunities*, Frances Pinter Centre of African Studies, Edinburgh, pp. 111–128.

Bourne, L.T., Lambert, E.V. and Steyn, K. (2002) Where Does the Black Population of South Africa Stand on the Nutrition Transition? *Public Health Nutrition*, 5(1a): 157–62.

Breisinger, C., Ecker, O., Maystadt, J.-F., Tan, J.-F.T., Al-Riffai, P., Bouzar, K., Sma, A. and Abdelgadir, M. (2014) *How to Build Resilience to Conflict: The Role of Food Security*, International Food Policy Research Institute, Washington, DC.

Breusers, M., Nederlof, S. and van Rheenen, T. (1998) Conflict or Symbiosis? Disentangling Farmer Herdsman Relations: The Mossi and Fulbe of the Central Plateau, Burkina Faso, *Journal of Modern African Studies*, 36: 357–380.

Brinkman, H.J. and Hendrix, C.S. (2011) Food Insecurity and Violent Conflict Causes, Consequences and Addressing the Challenges, World Food Programme Occasional Paper 24, World Food Programme, Rome.

Brottem, L.V. (2016) Environmental Change and Farmer–Herder Conflict in Agro-Pastoral West Africa, *Human Ecology*, 44(5): 547–563.

Bruijn, M. de, and Van Dijk, H. (2003) Changing Population Mobility in West Africa: Fulbe Pastoralists in Central and South Mali, *African Affairs*, 102(407): 285–307.

Bundervoet, T., Verwimp, P. and Akresh, R. (2009) Health and Civil War in Rural Burundi, *Journal of Human Resources*, 44(2): 536–563.

Cabot, C. (2017) *Climate Change, Security Risks and Conflict Reduction in Africa. Hexagon Series on Human and Environmental Security and Peace: A Case Study of Farmer–Herder Conflicts over Natural Resources in Côte d'Ivoire, Ghana and Burkina 1960–2000*, Springer-Verlag, Berlin and Heidelberg.

Cadre Harmonisé (2018) Report for Identifying Risk Areas and Vulnerable Populations in Sixteen (16) States and the Federal Capital Territory of Nigeria – March 2018. Published at: https://reliefweb.int/sites/reliefweb.int/files/resources/fiche-communic_march_2018-ch_23-3-2018.pdf. Accessed on 30 September 2018.

Carlson, J. and Chappell, M.J. (2015) Deepening Food Democracy, Institute For Agriculture and Trade Policy (IATP). Published at: http://safsf.org/documents/repository/220_04-30-15_2015_01_06_Agrodemocracy_JC_JC_f_0.pdf. Accessed on 6 April 2018.

Central Intelligence Agency (2018) *CIA World Factbook*. Published at: https://www. cia.gov/library/publications/the-world-factbook. Accessed on 16 March 2018.

Chauvin, N.D., Mulangu, F. and Porto, G. (2012) Food Production and Consumption Trends in Sub-Saharan Africa: Prospects for the Transformation of the Agriculture Sector, United Nations Development Programme Working Paper 2012-011.

Codjoe, S.N.A., Okutu, D. and Abu, M. (2016) Urban Household Characteristics and Dietary Diversity: An Analysis of Food Security in Accra, Ghana, *Food and Nutrition Bulletin*, 37(2): 202–218.

Collinson, P. and Macbeth, H. (2014) Introduction. In Collinson, P. and Macbeth, H. (eds) *Food in Zones of Conflict: Cross-Disciplinary Perspectives*, Berghahn Books, New York and Oxford.

Conceição, P., Levine, S., Lipton, M. and Warren-Rodríguez, A. (2016) Toward a Food Secure Future: Ensuring Food Security for Sustainable Human Development in Sub-Saharan Africa, *Food Policy*, 60: 1–9.

Conciliation Resources (2015) *Analysis of Conflict and Peacebuilding in the Central African Republic*, London.

Connolly-Boutin, L. and Smit, B. (2016) Climate Change, Food Security, and Livelihoods in Sub-Saharan Africa, *Regional Environmental Change*, 16(2): 385–399.

Cornia, G.A., van der Hoeven, R. and Mkandawire, T. (1992) The Supply Side: Changing Production Structures and Accelerating Growth. In Cornia, G.A, van der Hoeven, R. and Mkandawire, T. (eds) *Africa's Recovery in the 1990s: From Stagnation and Adjustment to Human Development*, Macmillan, Basingstoke.

Dawson, N., Martin, A. and Sikor, T. (2016) Green Revolution in Sub-Saharan Africa: Implications of Imposed Innovation for the Wellbeing of Rural Smallholders, *World Development*, 78: 204–18.

De Haan, C., Dubern, E., Garancher, B. and Quintero, C. (2016) *Pastoralism Development in the Sahel: A Road to Stability?* The World Bank, Washington, DC.

Deininger, K. and Castagnini, R. (2006) Incidence and Impact of Local Conflict in Uganda, *Journal of Economic Behaviour and Organisation*, 60(3): 321–45.

Devereux, S. (2016) Drivers of Household Food Availability in Sub-Saharan Africa based on Big Data from Small Farms, *Food Policy*, 60: 52–62.

Djurfeldt, G., Holmén, H., Jirström, M., and Larsson, R. (2005) The African Food Crisis: Lessons from Asian Experiences. In Djurfeldt, G., Holmén, H., Jirström, M., and Larsson, R. (eds) *The Africa Food Crisis: Lessons from the Asian Green Revolution*, CABI, Wallingford, UK / Cambridge, MA, pp. 1–16.

Djurfeldt, G., Holmén, H., Jirström, M. and Larsson, R. (2006) *Addressing Food Crisis in Africa: What Can sub-Saharan Africa Learn from Asian Experiences in Addressing its Food Crisis?* Swedish International Development Cooperation Agency (Sida) / Lund University, Lund, Sweden.

Ecker, O. (2014) Resilience for Food Security in the Face of Civil Conflict in Yemen. In Fan, S., Pandya-Lorch, R. and Sivan, Y. (eds) *Resilience for Food and Nutrition Security*, International Food Policy Research Institute (IFPRI), Washington, DC, pp. 53–63.

Ejigu, M. (2009) Environmental Scarcity, Insecurity and Conflict: The Cases of Uganda, Rwanda, Ethiopia and Burundi. In Brauch, H.G., Oswald Spring, U., Mesjasz, C., Grin, J., Kameri-Mbote, P., Chourou, B., Dunay, P. and Birkmann, J. (eds) *Coping with Global Environmental Change, Disasters and Security: Threats, Challenges, Vulnerabilities and Risks.* Hexagon Series on Human and Environmental Security and Peace, vol. 5. Springer, Berlin–Heidelberg–New York, pp. 719–733.

Evans-Pritchard, E.E. ([1948] 2014) *The Divine Kingship of the Shilluk of the Nilotic Sudan*, Cambridge University Press, Cambridge.

Food and Agriculture Organization (2015) *Africa Regional Overview of Food Insecurity 2015*, Accra, Ghana. Published at: http://www.fao.org/3/a-i4635e.pdf. Accessed on 15 May 2016.

Food and Agriculture Organization (2017a) *Food and Agriculture Organization Basic Definitions of Hunger.* Published at: http://ecsw.org/files/global/world-hunger/news/fao-basic-definitions-of-hunger.pdf. Accessed on 7 April 2018.

Food and Agriculture Organization (2017b) *Africa Regional Overview of Food Security and Nutrition 2017*, Accra, Ghana. Published at: http://www.fao.org/publications/card/en/c/a0745c69-2412-4f56-afa9-6b011d2e0b71/. Accessed on 7 April 2018.

Food and Agriculture Organization (FAO), IFAD, UNICEF, World Food Programme and WHO (2017) *The State of Food Security and Nutrition in the World 2017*, Food and Agriculture Organization, Rome. Published at: http://www.fao.org/3/a-I7695e.pdf, fao.org/3/a-I7695e.pdf. Accessed on 12 April 2018.

Fjelde, H. and Uexkull, N. (2012) Climate Triggers: Rainfall Anomalies, Vulnerability and Communal Conflict in Sub-Saharan Africa, *Political Geography*, 31(7): 444–453.

Frelat, R., Lopez-Ridaura, S., Giller, K.E., Herrero, M., Douxchamps, S, Andersson, A.D., Erenstein, O., Henderson, B., Kassie, M., Birthe, K.P., Rigolot, C., Ritzema, R.R., Rodriguez, D., van Asten, P.J.A, and van Wijk, M.T. (2016) Drivers of household food availability in sub-Saharan Africa based on big data from small farms *Proceedings of the National Academy of Sciences of the United States of America*, 113(2): pp. 458–463.

Global Harvest Initiative (2012) *2012 GAP Report: Measuring Global Agricultural Productivity.* Published at: http://www.globalharvestinitiative.org/GAP/GHI_2012_GAP_Report_Int.pdf. Accessed on 6 April 2018.

Grebmer, K. von, Bernstein, J., Brown, T., Prasai, P., Yohannes, Y., Towey, O., Foley, C., Patterson, F., Sonntag, A. and Zimmermann, S.-M. (2017) *2017 Global Hunger Index: The Inequalities of Hunger*, International Food Policy Research Institute, Washington, DC.

Guardian (2013) How Africa Can Solve its Food Crisis by Growing More Crops Sustainably. Published at: https://www.theguardian.com/global-development/poverty-matters/2013/apr/18/africa-food-crisis-growing-crops-sustainably. Accessed on 30 March 2018.

Guardian (2017) Uganda at Breaking Point: Bidi Bidi Becomes World's Largest Refugee Camp. Published at: https://www.theguardian.com/global-development/2017/apr/03/uganda-at-breaking-point-bidi-bidi-becomes-worlds-largest-refugee-camp-south-sudan. Accessed on 10 June 2017.

Hardin, G. (1968) The Tragedy of the Commons, *Science*, 162(3859): 1243–1248.

International Crisis Group (2017) *Herders against Farmers: Nigeria's Expanding Deadly Conflict.* Report 252, 19 September 2017. Published at: https://www.crisisgroup.org/africa/west-africa/nigeria/252-herders-against-farmers-nigerias-expanding-deadly-conflict. Accessed on 18 April 2018.

Internal Displacement Monitoring Centre (IDMC) (2016) Displacement Caused by Conflict and Disasters 2015. Published at http://www.internal-displacement.org/assets/publications/2016/2016-global-report-internal-displacement-IDMC.pdf. Accessed on 6 July 2016.

Internal Organization for Migration (IOM) (2018) Resurgence of Communal, Armed Conflict in Northern Mali Causes Population Displacement. Published at: https://www.iom.int/news/resurgence-communal-armed-conflict-northern-mali-causes-population-displacement. Accessed on 6 April 2018.

Intergovernmental Panel on Climate Change (IPCC) (2007) Front Matter. In Parry, M.L., Canziani, O.F., Palutikof, J.P., van der Linden, P.J. and Hanson, C.E. (eds) *Climate Change 2007: Impacts, Adaptation and Vulnerability: Contribution of Working Group II to the Fourth Assessment Report of the Intergovernmental Panel on Climate Change (SAR)*, Cambridge University Press, Cambridge and New York, pp. 1–6. Published at: http://www.ipcc.ch/publications_and_data/ar4/wg2/en/contents.html. Accessed on 31 March 2018.

IPC (2018). *Integrated Phase Classification Overview and Classification System*. Published at: http://www.ipcinfo.org/ipcinfo-website/ipc-overview-and-classification-system/en/. Accessed on 30 September 2018.

Ittersum M.K. van, van Bussel, L.G.J., Wolf, J., Grassini, P., van Wart, J., Guilpart, N., Claessens, L., de Groot, H., Wiebe, K., Mason-D'Croz, D., Yang, S.H., Boogaard, H., van Oort, P.A.J., van Loon, M., Saito, K., Adimo, O., Adjei-Nsiah, S., Alhassane, A., Bala, A., Chikowo, R., Kaizzi, K.C., Kouressy, M., Makoi, J.H.J.R., Ouattara, K., Tesfaye, K. and Cassman, K.G. (2016) Can Sub-Saharan Africa Feed Itself? *Proceedings of the National Academy of Sciences* (PNAS), 113: 14964–14969.

Kasozi, J. (2017) The Refugee Crisis and the Situation in Sub-Saharan Africa, Österreichische Gesellschaft für Europapolitik Policy Brief, 16.

Kinyoki, D.K., Moloney, G.M., Uthman, O.A., Kandala, N-B., Odundo, E.O., Noor, A.M. and Berkley, J.A. (2017) Conflict in Somalia: Impact on Child Undernutrition, *British Medical Journal Global Health*, 2(2): 1–11.

Koren, O. and Bagozzi, B.E. (2017) Living Off the Land: The Connection Between Cropland, Food Security and Violence Against Civilians, *Journal of Peace Research* 54(3): 351–364.

Lienhardt, G. (1954) The Shilluk of the Upper Nile. In Forde, D. (ed.) *African Worlds, Studies in the Cosmological Ideas and Social Values of African Peoples,* International African Institute, Oxford University Press, Oxford, pp. 138–164.

Luan, Y., Cui, X. and Ferrat, M. (2013) Historical Trends of Food Self-Sufficiency in Agriculture, *Food Security*, 5(3): 393–405.

Maunder, N., and Wiggins, S. (2007) Food Security in Southern Africa: Changing the Trend? Review of Lessons Learnt on Recent Responses to Chronic and Transitory Hunger and Vulnerability, *Natural Resources Perspectives*, 106 (June), Overseas Development Institute, London.

Mazo, J. (2010) *Climate Conflict: How Global Warming Threatens Security and What to Do About It*, Routledge, New York.

Minoiu, C. and Shemyakina, O. (2012) Child Health and Conflict in Côte D'Ivoire. *American Economic Review: Papers and Proceedings*, 102(3): 294–299.

Moritz, M. (2010) Understanding Herder–Farmer Conflicts in West Africa: Outline of a Processual Approach, *Human Organisation*, 69(2): 138–148.

Munang, R. and Andrews, J. (2014) Despite Climate Change, Africa Can Feed Itself. In Tafirenyika, M. (ed.) *Africa Renewal 2014 Special Edition*, United Nations, New York.

Nori, M., Switzer, J. and Crawford, A. (2005) *Herding on the Brink: Towards a Global Survey of Pastoral Communities and Conflict*, International Institute for Sustainable Development, Winnipeg.

Norwood, F.B. (2015) Understanding the Food Democracy Movement, *Choices. The Magazine of Food, Farm and Resource Issues*, 30(4): 1–5.

OECD–Food and Agriculture Organization (2018) *OECD–Food and Agriculture Organization Agricultural Outlook 2018–2027*, OECD, Paris / Food and Agriculture Organization, Rome.

Ofuoku, A.U. and Isife, B.I. (2009) Causes, Effects and Resolution of Farmer–Nomadic Cattle Herders Conflict in Delta State, Nigeria, *International Journal of Sociology and Anthropology*, 1: 47–54.

Ohlsson, L. (1999) *Environment, Scarcity and Conflict: A Study of Malthusian Concerns*, Department of Peace and Development Research, Göteborg Universität, Gothenburg.

Olaniyan, A. (2015) The Fulani–Konkomba Conflict and Management Strategy in Gushiegu, Ghana, *Journal of Applied Security Research*, 10(3): 330–340.

Oyeniyi, B.A. and Akinyoade, A. (2014) Landmines, Cluster Bombs and Food Insecurity in Africa. In Collinson, P. and Macbeth, H. (eds) *Food in Zones of Conflict: Cross-Disciplinary Perspectives*, Berghahn Books, New York and Oxford.

Pinstrup-Andersen, P. and Shimokawa, S. (2008) Do Poverty and Poor Health and Nutrition Increase the Risk of Armed Conflict Onset? *Food Policy*, 33(6): 513–520.

Rakotoarisoa, M.A., Iafrate, M. and Pascali, M. (2011) *Why has Africa Become a Net Food Importer? Explaining Africa Agricultural and Trade Deficits*, Food and Agriculture Organization, Rome.

Reynolds, T.W., Waddington, S.R., Anderson, C. L., Chew, A., True, Z. and Cullen, A. (2015) Environmental Impacts and Constraints Associated with the Production of Major Food Crops in Sub-Saharan Africa and South Asia, *Food Security*, 7(4): 795–822.

Schilling, J.P., Scheffran, J. and Link, M. (2010) Climate Change and Land Use Conflicts in Northern Africa, *Nova Acta Leopoldina*, 112(384): 173–182.

Sen, A. (1981) *Poverty and Famines: An Essay on Entitlement and Deprivation*. Oxford University Press, Oxford.

Shettima, A.G. and Tar, U.A. (2008) Farmer–Pastoralist Conflict in West Africa: Exploring the Causes and Consequences, *Information, Society and Justice*, 1(2): 163–184.

The Economist (2017) (9 November) African Herders Have Been Pushed into Destitution and Crime. Published at: https://www.economist.com/news/middle-east-and-africa/21731191-owning-cattle-excellent-way-hide-ill-gotten-wealth-too-afri-can-herders. Accessed on 10 February 2018.

Toenniessen, G., Adesina, A. and Devries, J. (2008) Building an Alliance for a Green Revolution in Africa, *Annals of the New York Academy of Sciences*, 1136: 233–242.

Tonah, S. (2006) Migration and Farmer–Herder Conflicts in Ghana's Volta Basin. *Canadian Journal of African Studies / Revue Canadienne Des Études Africaines*, 40(1): 152–178.

Transparency International (2018) Corruption Perceptions Index 2017. Published at: https://www.transparency.org/news/feature/corruption_perceptions_index_2017. Accessed on 1 March 2018.

United Nations, Department of Economic and Social Affairs, Population Division (2017) *World Population Prospects: The 2017 Revision, Key Findings and Advance Tables.* Working Paper No. ESA/P/WP/248.

United Nations Development Programme (UNDP) (2017) Global 2016 Human Development Report. Published at: http://hdr.undp.org/sites/default/files/2016_human_development_report.pdf. Accessed on 7 April 2018.

United Nations Economic and Social Council: United Nations; Economic Commission for Africa (2017) Africa Review Report on Drought and Desertification. Published at: https://repository.uneca.org/handle/10855/3694. Accessed on 31 March 2018.

UNHCR (2018) Operational Update Mali January 2018. Published at: https://reliefweb.int/sites/reliefweb.int/files/resources/62182.pdf. Accessed on 30 March 2018.

UNOCHA (2018a) Sahel 2018. Overview of Humanitarian Needs and Requirements. Published at: https://reliefweb.int/report/nigeria/sahel-2018-overview-humanitarian-needs-and-requirements. Accessed on 1 April 2018.

UNOCHA (2018b). Chad. Food Security and Nutrition Crisis. Appeal for a Response at Scale. Published at: https://reliefweb.int/sites/reliefweb.int/files/resources/tcd_advo_food_security_and_nutrition_crisis_english3.pdf. Accessed on 30 September 2018.

UNOCHA (2018c) Sudan. Humanitarian Needs Overview. February 2018. Published at: https://reliefweb.int/sites/reliefweb.int/files/resources/Sudan_2018_Humanitarian_Needs_Overview.pdf. Accessed on 1 April 2018.

UNOCHA (2018d) Mali. Humanitarian Bulletin January–February 2018. Published at: https://reliefweb.int/sites/reliefweb.int/files/resources/MLI_BIH_Jan%20Feb%20 2018_Eng_final.pdf. Accessed on 1 April 2018.

UNOCHA (2018e) South Sudan Humanitarian Snapshot, 12 January 2018. Published at: https://reliefweb.int/report/south-sudan/south-sudan-humanitarian-snapshot-december-2017. Accessed on 7 April 2018.

USAID (2017) Food Assistant Factsheet December 2017. Published at: https://www.usaid.gov/central-african-republic/food-assistance. Accessed on 1 April 2018.

USAID (2018) Food Assistance Fact Sheet – Cameroon. 4 September 2018. Published at: https://www.usaid.gov/cameroon/food-assistance. Accessed on 30 September 2018.

Vanlauwea, B., Wendtb, J., Gillerc, K.E., Corbeelsde, M., Gerardf, B. and Nolteg, C. (2014) A Fourth Principle is Required to Define Conservation Agriculture in Sub-Saharan Africa: The Appropriate Use of Fertilizer to Enhance Crop Productivity, *Field Crops Research*, 155: 10–13.

World Bank (2014) World Bank Boosts Support for Pastoralists in Horn of Africa. Press release, 18 March 2014. Published at: http://www.worldbank.org/en/news/press-release/2014/03/18/world-bank-pastoralists-horn-africa. Accessed on 31 March 2018.

World Food Programme (2018a) World Food Programme Senegal Country Brief January 2018. Published at https://reliefweb.int/sites/reliefweb.int/files/resources/Senegal%20CB%20Jan%202018-RBD.pdf. Accessed on 1 April 2018.

World Food Programme (2018b) World Food Programme Somalia. Published at: http://www1.wfp.org/countries/somalia. Accessed on 1 April 2018.

World Food Programme (2018c) World Food Programme South Sudan Situation
    Report No. 216, 29 March 2018. Published at: https://reliefweb.int/report/south-
    sudan/wfp-south-sudan-situation-report-216-29-march-2018. Accessed on 6 April
    2018.

CHAPTER 3
FROM HEALTHY TO SUSTAINABLE: TRANSFORMING THE CONCEPT OF THE MEDITERRANEAN DIET FROM HEALTH TO SUSTAINABILITY THROUGH CULTURE

F. Xavier Medina

Introduction

During recent years we have increasingly assisted in a debate on sustainability, food security and nutrition under the auspices of the Food and Agriculture Organization (FAO). In this framework, sustainable diets have emerged as a public health and nutrition challenge, as well as a critical issue for sustainable food systems (Burlinghame and Dernini 2012; Lang and Barling 2013; Johnston et al. 2014; Berry et al. 2015; Meybek 2015; Dernini et al. 2017).

The concept of sustainable diets acknowledges the interdependencies of food production and consumption with food requirements and nutrient recommendations, and at the same time reaffirms the notion that the health of humans cannot be isolated from that of ecosystems (Burlinghame and Dernini 2012). As Dernini et al. point out: 'Sustainable diets, which are ecosystem specific, are but one practical way of applying sustainability to food security and nutrition. In this overall context, sustainability becomes the long-term component of all the levels and dimensions of food security – the well-established and accepted determinant of a nation's health and well-being' (Dernini et al. 2017: 1).

The incorporation of sustainability issues into dietary guidelines has been increasingly discussed since the 1980s to make them healthier for the environment as well as for the consumers. In a well-known paper, Gussow and Clancy (1986) suggested the term 'sustainable diet' to describe a diet based on food that is chosen not only regarding health, but also regarding sustainability (that is, a certain capability of food maintenance into the foreseeable future). They concluded that consumers should, when possible, buy 'locally produced foods', thus making it less expensive in energy terms (due to less transport) while also supporting the local and regional agriculture.

Today, however, a main concern seems to be to conserve natural resources for future generations, while simultaneously providing enough food, in quantity and quality, to meet the nutritional requirements of a growing global population. Within this framework, the interest in the Mediterranean diet as a model of a sustainable dietary pattern has increased (Gussow 1995; Burlinghame and Dernini 2011; Dernini and Berry 2015).

This chapter aims to conduct a conceptual and diachronic review on the construction of the Mediterranean diet as a subject of analysis. The concept of the Mediterranean diet came into being shortly after the mid-twentieth century as a recommended and healthy diet, mainly aimed at North American society. Since then, it has undergone various modifications that have led it from being a concept linked solely to health, to an element of culture and a lifestyle, as a result of its declaration as an 'Intangible Cultural Heritage' (ICH) by UNESCO in 2010 (Medina 2009, 2011; Serra Majem and Medina 2015). Since that point, the Mediterranean diet has been set along a new path, guided by the FAO, as a sustainable diet, and having sustainability and locality as the cornerstones of its new identity. A historic transformation of the concept of the Mediterranean diet is therefore apparent, which has taken it from health to sustainability, through culture. However, it is also evident that the focus on health has never really declined in importance, but has instead been adapted to the times, and modified to suit new food trends.

A Mediterranean Framework

Like so many regions, the 'Mediterranean' is a sociocultural and political construct based on a geographical entity. Any discussion of food in the Mediterranean also involves a social and cultural construction. Food, and eating behaviours in general, fall within the framework of the societies that produce and recreate them, and therefore within specific sociocultural systems (Medina 1996). Whilst we construct our conception of the Mediterranean based on certain parameters, which are strongly defined by geography and climate, they are also defined by cultural projections and stereotypes that are difficult to avoid. We use these to create differences and similarities, and we also define them in relation to agriculture, livestock farming and food. This means that it is possible to say that despite being geographically close, certain cultures and/or religions, such as those of the populations north and south of the Mediterranean, are very different and even conflicting; on the other hand, some of them are not so distinct and even have a type of family relationship, which gives them their identity and ultimately brings them together. In this regard, the Spanish anthropologist Julio Caro Baroja quoted Diego de Torres, the first historian of Morocco's sharifs in the sixteenth century, who had compared the Moroccan villages in the Atlas Mountains with the Andalusian villages in the Alpujarra region:

> In both the physical environment and the social environment, to our eyes they appear to be very similar to what one might see in the Alpujarra region of the Sierra Nevada. The villages located on the northern slopes of the Atlas were abundant in water; their fields were full of orange and lemon trees in the lower and more sheltered parts, and they had ample grain grown on the slopes, as well as fruit trees of a European type. (Caro Baroja 1956: 56)

A great deal has been written regarding the definition of what the Mediterranean 'is' or 'is not'. The French anthropologist, Igor de Garine, like Fernand Braudel (1981) before him, in one of his most classic writings, tells us:

> Perhaps the distribution of the olive grove marks the interior limits of what we believe to be the Mediterranean agricultural world. There are numerous variations of nuances, from north to south and from east to west, but we can comfortably define the Mediterranean area as the one that enables cereals, vines and olive groves to be cultivated without any risk, while engaging in livestock farming, formerly nomadic, now sedentary, in which animals of ovine and caprine species predominate. (Garine 1993: 11)

However, the Mediterranean is and has always been a place of communication and contact. As Fernand Braudel noted (1981), the routes in and around this 'Inland Sea' can be found on land and at sea and are linked in a system of circulation that has driven all types of relationships between the various societies that have developed on its shores. The Mediterranean is therefore, on the one hand, an important crossroads and a vast centre for integrating and redistributing influences – not only in food – received from the four cardinal points; and, on the other, an exceptional field of acclimatisation, where products, most of which still exist today, from other places both near and far, have found a suitable environment for their cultivation or breeding, and thereby expect to be accepted and implemented within the local food systems.

But this ability to integrate influences is no coincidence. It should be remembered that many of our cuisines are the product of processes rather than recipes. They thus have a gift for adopting new raw materials to the methods that they have always used. For example, as observed by González

Turmo (1993), over time new vegetables or meats have been introduced to many Andalusian stews, without changing the essentials of the cuisine.

Food Culture and Heritage in the Mediterranean Area

Since the turn of the last century, with the UNESCO 'Proclamation of Masterpieces of the Oral and Intangible Heritage of Humanity' international distinction, and subsequently with the UNESCO declaration of the first spaces that constituted intangible heritage of humanity, the 'official' concept of heritage can be considered to have begun to take an interest in areas beyond the purely monumental and environmental fields, broadening its scope to more ethno-anthropological and less tangible aspects.

An important point to emphasise here is that gastronomic heritage, and therefore human consumption in general, falls within this emerging intangible heritage. If the UNESCO heritage declaration of 2005 had an outstanding characteristic (from the point of view of this study) it was that, for the first time, a country like Mexico was presenting its culinary art at a national level, in order for it to be declared world heritage. The candidacy was rejected, but Mexico announced that it would submit it once again, as an essential part of its culture. In order to highlight the heritage value of its cuisine, in July 2008 the Mexican candidacy convened an international academic meeting in the city of Campeche (Mexico) entitled: 'The cuisine as cultural heritage – criteria and definitions'; its aim was to provide UNESCO with a series of recommendations that would lead to greater awareness of food/gastronomic candidacies (known as the Declaration of Campeche, 2008). A similar initiative took place a year later in Barcelona, to request candidacy of the Mediterranean diet (the Barcelona Declaration in 2009). However, the zenith of UNESCO recognition for food candidacies did not arrive until November 2010, when the three proposals submitted at that time were declared Intangible Cultural Heritage: traditional Mexican cuisine, the gastronomic meal of the French (*le Répas gastronomique des Français*) and the Mediterranean diet. Food had, for the first time, been recognised by UNESCO as Intangible Cultural Heritage of Humanity.[1]

On the following pages, I shall discuss in more specific terms the construction of the candidacy of the Mediterranean diet as Intangible Cultural Heritage, and its path towards the final declaration in 2010. As with any UNESCO candidacy, the proposal involved negotiations, teams, conceptual apparatus, bureaucracies, times and economies, which when observed as a whole and at a distance, are highly illustrative of how food heritage is constructed and negotiated as cultural heritage and in relation to a supranational institution like UNESCO.

The Conception of the Candidacy and Debates on the Concept of the Mediterranean Diet

In 2004, in the wake of the application to UNESCO for Mexican cuisine to be recognised, the idea of presenting the Mediterranean diet as Intangible Cultural Heritage emerged. The proposal was made by the Mediterranean Diet Foundation (MDF), which was based in Barcelona and was at that time chaired by Lluís Serra Majem. The Spanish Ministry of Agriculture, Fisheries and Food quickly lent its support to the candidacy, and played a key role at all levels, especially financial.[2]

The first international preparatory meeting to launch the bid took place in Rome in 2005 (declared Year of the Mediterranean by the European Commission) during the Third Euro-Mediterranean Forum 'Dialogues between Civilizations and Peoples of the Mediterranean: The Food Cultures'. As this was the first public international event for the future candidacy, the organisers proposed to include a round-table discussion on the strategic need for a consensus on the Mediterranean diet, considered in terms of a cultural system as well as a medical diet. To that end, an interdisciplinary debate took place on the definition of the Mediterranean diet.

According to the general coordinator of the meeting Sandro Dernini (2008) (and subsequent members of the Italian team responsible for the final wording of the application):

> The Mediterranean diet as a whole lifestyle makes the cultural identities and diversity visible, providing a direct measure of the vitality of the culture in which it is embedded. The Mediterranean diet is an expression of a Mediterranean style of life in continuous evolution throughout time, and it is constantly recreated by communities and groups in response to changes in their environment and history. As a part of the different cultures, it provides a sense of identity and continuity for the Mediterranean societies.

In October 2009, the International Scientific Committee of the MDF met in Barcelona and adopted the Barcelona Declaration on the Mediterranean diet as an Intangible Cultural Heritage. In December of the same year, the Spanish Ministry of Agriculture, Fisheries and Food held the first international institutional meeting in Madrid, to which their counterparts in Greece, Italy and Morocco were also invited, and at which the preparation of a joint transnational candidacy was agreed. In the following spring, at the headquarters of Italy's Ministry of Agriculture, Food and Forestry Policies in Rome, the four states formalised the process in the Rome Declaration, and the MDF was designated as the Transnational Technical Coordinator of the application.

The Mediterranean Diet Foundation has since maintained that:

The ancient Greek word *diaita* (δίαιτα), from which diet is derived, means a balanced lifestyle, and this is exactly what the Mediterranean diet is – much more than a nutritional pattern. The Mediterranean diet is a way of life, not just a food pattern that combines ingredients from the local agriculture, recipes and cooking methods of each place, shared meals, celebrations and traditions, which together with moderate but daily physical exercise, favoured by a mild climate, complete the lifestyle that modern science invites us to adopt for the benefit of our health, making it an excellent model for healthy living.[3]

One of the major problems during the initial management of the candidacy was the name of the heritage item, whether 'Mediterranean Diet' or 'Mediterranean Food'. Whereas those involved in the (very powerful) spheres of public health and even agriculture strongly favoured use of the well-established and well-known term, 'Mediterranean Diet', those advocating perspectives more closely linked to the social and cultural sphere argued for use of the term 'Mediterranean Food', as being broader, more inclusive and, above all, less ideologically linked to an established medical concept. Then, within anthropology, a debate was sparked between the two concepts and the choice of term. Finally, the term 'Mediterranean Diet' won out.

Despite this, the definition of the Mediterranean diet in the candidacy obviously attempted to be broad and inclusive. According to the text presenting the candidacy, 'The Mediterranean diet is an articulated cultural ensemble manifested in the following domains specified in the Convention (Article 2.2)', which means that it represents the domains of oral traditions and expressions, ritual social practices and feasts, the knowledge and practices related to nature and the universe, and traditional crafts (UNESCO 2010).

It is important to stress that the definition always refers to the entire concept of the Mediterranean diet, and not to each of its possible tangible or intangible components individually, such as olive oil, specific dishes, landscapes, festivals, rituals, etc. However, they are all individual examples of the substantial components of the item inscribed. In other words, they are a necessary constituent part of the Mediterranean diet and are part of the Intangible Cultural Heritage of Humanity, although none of them has been recognised as such on an individual basis. Furthermore, those components that may or may not form part of the heritage as inscribed cannot be selected arbitrarily on the basis of a partial criterion (such as whether they are healthy). This is a key aspect to take into account.

Another important aspect that must be borne in mind is that the Mediterranean diet is an item of cultural heritage, and, as such, it is living and constantly changing. In other words, it is made up of ductile sociocultural structures and elements, which evolve as we and our societies evolve. We cannot, therefore, hope for this cultural heritage to remain unchanged over time (even when declaring something to be an ICH and demanding its protection and safekeeping), and neither can we hope for it to remain isolated from the various influences that may partially or even substantially change it. This capacity for evolution is intrinsic to its cultural constitution, and as such should be accepted.

There is therefore no point in attempting to conceive of an item of cultural heritage based on specific fixed and non-negotiable parameters, which are used as the basis for its definition – as has been attempted on several occasions – or expecting to declare a Mediterranean diet based on the rural diet prior to the Second World War as Intangible Cultural Heritage, and defining traditional products based on the same premise (Trichopoulou 2012: 951). The primary consideration in the inscription of an item as world heritage is that the item exists, is alive and still in use. Otherwise, we would be attempting to register something historical, which would refer more to the memory of the past than to everyday use.

All these aspects (some still not entirely resolved) became apparent after 2008, when the ministries of agriculture and/or health of the four countries proposing the candidacy had to step aside and pass the baton to their ministries of culture, which are now legally and administratively responsible for presenting proposals of this type to UNESCO.

After difficult transnational negotiations between the four member countries of the candidacy, it was presented and finally declared an ICH of humanity by UNESCO at the meeting of the Intergovernmental Committee, which took place in Nairobi, Kenya on 16 November 2010. At the same meeting, Mexican cuisine and French gastronomy were also each declared an International Cultural Heritage, which meant that three directly food-related items (which were not cultural landscapes) were on the UNESCO heritage list for the first time.

It is important to note that 2013 saw the adherence to the Mediterranean diet of three new countries: Cyprus, Croatia and Portugal. Since that year, there have therefore been seven countries that officially consider this heritage as their own. Further affiliations are also possible, meaning that other countries may decide to join in the future.

Heritage and Its Safeguarding

The significance of this declaration lies in the fact that intangible cultural assets related to food have been declared human heritage by a supranational institution for the first time in history. The declaration is also, and above all, a recognition of the urgent need to preserve the techniques, practices, habits, ideas, values and spaces of the food cultures approved for inclusion.

Their safeguarding directly affects the communities and groups which are the basis for this heritage and which maintain it, its ways of life and social organisation. This is no trivial matter, and as we noted a few years ago in a critical article (González Turmo and Medina 2012) specifically about the Mediterranean diet, this safeguarding is complex, and risks removing the issue in question from its context.

Given the scale of the heritage area proposed in the candidacy approved by UNESCO, it is clear that the governments and institutions in the four countries that led it (and the three that subsequently joined it) have assumed a major responsibility, which we are not sure they have understood to its full extent. It is no simple challenge: the food culture is a complex one, which cannot be defined by merely listing the foodstuffs within it, its culinary preparations or its most distinctive rituals – as some governments and individuals might prefer. The same can be said of its relationship to health, which, although no one disputes a certain connectedness, is only a small part of a much more complex whole. Furthermore, this was not the aim of the candidacy.

Many important challenges were assumed when submitting the candidacies, due to the urgent need to safeguard the heritage values that, according to the dossier, affect the Mediterranean peoples and their cultures, their cultural spaces, identity and intercultural dialogue, knowledge and creativity. The initiative now lies with the governments that promoted it; they must not now shirk a responsibility that demands urgent and necessary action.

We must acknowledge that the declaration of the Mediterranean diet as heritage has taken place against the backdrop of an economic downturn. The crisis that began around 2007 has been directly affecting the countries concerned: Greece, Italy, Portugal, Spain and Cyprus. Nevertheless, budget cuts cannot and should not be arguments to justify inaction, but instead should force us to find new ways to value something that is as important to humans as food.

Mediterranean Diet 4.0: The Sustainable Mediterranean Diet

As Dernini (2008) points out, the Mediterranean diet is actually observed as a whole lifestyle that makes the cultural identities and diversity visible, providing a direct measure of the vitality of the culture in which it is embedded. In this sense, the Mediterranean diet is an expression of a Mediterranean style of life in continuous evolution through time, and it is slowly but constantly recreated by communities and groups in response to changes in their environment and history. As a part of the different cultures, the Mediterranean diet provides a sense of identity and continuity for Mediterranean societies.

Furthermore, with the modernisation of agriculture and globalisation of foods that took place in the second half of the twentieth century, concepts such as sustainable diets and human ecology have been neglected in favour of intensification and industrialisation of agricultural systems. More recently, in the last three decades, the growing concern over food safety has motivated a renewed interest in organic foods (Herrin and Gussow 1989) and in locally produced and sustainable foods. This fact is particularly interesting in the Mediterranean area (Medina 2015), where international movements, such as Slow Food, are based on the defence of local production, biodiversity and sustainability, where both sociocultural and biological aspects are included.

Nevertheless, the Mediterranean diet is still being considered politically as an independent entity, and not as part of a significant social and cultural Mediterranean food system. Health or food consumption is still considered separately from agricultural and fisheries production, economics (sales, import–export, etc.) and the maintenance of traditional structures of distribution and sale (González Turmo and Medina 2012). In this framework, while good nutrition should be a goal of agriculture, it is imperative that concerns of sustainability are not lost in the process. Many dietary patterns can be healthy, but they can vary substantially – for example, in terms of their resource cost or their environmental impact.

In 2011, the Mediterranean diet was identified by FAO and the International Centre for Advanced Mediterranean Agronomic Studies (CIHEAM) as a joint case study for characterisation and assessment of the sustainability of dietary patterns in different agro-ecological zones, with specific regard to the sustainability of Mediterranean food systems. This led, in 2012, at the 9th CIHEAM meeting of the ministers of agriculture, to the acknowledgement of the role of the Mediterranean diet as 'a driver of sustainable food systems in the Mediterranean' (Dernini et al. 2017).

Under the auspices of FAO and CIHEAM, the Mediterranean diet is now going to be observed as a 'sustainable food system' in the Mediterranean area. The present framework includes the following points:

(1) nutritionally adequate, safe and healthy;
(2) low environmental impact – protective and respectful of biodiversity and ecosystem;
(3) culturally acceptable; and
(4) accessible, economically fair and affordable.

Within the three social, environmental and economic sustainability pillars, four mutually interdependent, thematic sustainability dimensions of the Mediterranean diet were also identified:

- nutrition and health
- environment including biodiversity
- sociocultural factors
- economy.

As an outcome of this interdisciplinary and multidimensional sustainability approach, the following four benefits were highlighted in one single comprehensive Med Diet 4.0 framework, with country specific variations:

- major health and nutrition outcomes
- low environmental impacts and rich in biodiversity
- high sociocultural food values
- positive economic return locally (Dernini et al. 2017)

This Med Diet 4.0 methodological approach tries to contribute to understanding what constitutes a sustainable diet in the different agro-ecological zones, by taking into account the specific country and regional variations within the Mediterranean area.

Conclusions

Like its cultures, the cuisines of the Mediterranean are different, diverse, interconnected and changing, continuously evolving, and receiving external influences that accompany their own internal developments. What is clear at all times is the difficulty of 'building' the idea of a common food (or a diet) for all the countries or cultures of the Mediterranean.

Attempting to find common elements between them is a difficult exercise (which is also a construction). Evidently, as it involves a shared space like the Mediterranean, with strong interactions and a long-shared history, it leads to some positive results. However, these results cannot be reductionist, and nor can we expect the different situations to fit our preconceived or desired models. It is, therefore, impossible to speak of a community of shared ingredients across all the countries in the Mediterranean basin, or of identical cuisines or dishes. However, we can speak primarily in terms of a community of common culinary techniques and preparations. This includes local adaptations that are the same in some cases; production and sales structures that have some shared characteristics; and even some forms of consumption, which when taken as a whole enable us to talk of a more or less common Mediterranean culinary system (Contreras et al. 2005; González Turmo 2005), and which confer some degree of family relationship on the various cuisines surrounding the Mediterranean, which has been called 'the Mediterranean diet' since its declaration as an Intangible Cultural Heritage by UNESCO in 2010.

Nevertheless, we must always bear in mind that the Mediterranean diet is part of an interdependent social and cultural system; it should never be considered a separate element in itself, as it sometimes is, especially by those involved in the health sphere and in relation to certain selected products. This food system is a complex network of interdependent cultural aspects and we must remember that all the links in the chain must be protected (Medina 2015) from production to the dish, including distribution, sales, cooking techniques, ingredients, consumer behaviour, etc.

What we think of today as the Mediterranean diet has undergone various changes that have taken it from a conception linked, at its origins and for some decades, solely to health, and made it an element of culture, a lifestyle, after its declaration as Intangible Cultural Heritage by UNESCO in 2010. Since that point, the Mediterranean diet has adopted a new path, guided by the FAO, that focuses on sustainability and locality as the cornerstones of its new identity. We can therefore see how the Mediterranean diet has undergone a conceptual transformation over the years, which has taken it from only health – a concept that, by the way, also has an important social component (Medina et al. 2014) – to include sustainability, by way of culture. It is also evident that, even if the focus on health has never really declined in importance, it has instead been adapted to the times and been modified to suit new food trends. Nevertheless, health remains its primary focus, and at present, the challenge is at the very core of the protection of food systems as cultural heritage.

F. Xavier Medina
Director of the UNESCO Chair in Food, Culture and Development, Faculty of Health Sciences, Universitat Oberta de Catalunya (UOC), Barcelona.

Notes

1. Nevertheless, it is true that both beforehand and afterwards, various initiatives and candidacies related (albeit tangentially) to the food sphere have been submitted (or are in the process of being submitted) to UNESCO. Those related to cultural landscapes include the jurisdiction and the landscape of the vineyards of Saint-Emilion (1999) in France, the wine region of the Alto Douro (2001) and the wine-producing landscape of the island of Pico (2004) in Portugal, the cultural landscape and the wine region of Tokaj in Hungary (2004), and the agave landscape and the ancient industrial facilities of tequila in Mexico (2006).
2. For this section, a reworking of various texts derived primarily from González Turmo and Medina (2012) and Serra Majem and Medina (2015) were used.
3. https://dietamediterranea.com/en/ (last accessed 14 February 2019).

References

Berry, E.M., Dernini, S. Burlingame, B., Meybeck, A. and Conforti, P. (2015) Food Security and Sustainability: Can One Exist without the Other? *Public Health Nutrition*, 16: 1–10.

Braudel, F. ([1949] 1981) *El Mediterráneo y el mundo mediterráneo en la época de Felipe II*, Fondo de Cultura Económica, Mexico.

Burlingame, B. and Dernini, S. (2011) Sustainable Diets: The Mediterranean Diet Example, *Public Health Nutrition*, 14(12A): 2285–2287.

Burlingame, B. and Dernini, S. (eds) (2012) *Sustainable Diets: Directions and Solutions for Policy, Research and Action*, Food and Agriculture Organization, Rome.

Cabo, E. de. (2013) Diez años de la convención sobre patrimonio inmaterial confirman su validez y su necesaria revisión constante, *Ph (en línea)*, 84 (October). Published at: http://www.iaph.es/revistaph/index.php/revistaph/article/view/3388. Accessed on 22 October 2017.

Caro Baroja, J. (1956) *Una visión de Marruecos a mediados del siglo XVI, la del primer historiador de los 'Xarifes', Diego de Torres*, Consejo Superior de Investigaciones Científicas, Madrid.

Contreras, J., Riera, A. and Medina, F. X. (2005) Introducción. In Contreras, J., Riera, A. and Medina, F. X. (eds) *Sabores del Mediterráneo: Aportaciones para promover un patrimonio alimentario común*, Institut Europeu de la Mediterrània, Barcelona.

Dernini, S. (2008) The Strategic Proposal to Candidate the Mediterranean Diet for Inscription in the UNESCO List of Intangible Cultural Heritage, *Med2008. Mediterranean Yearbook. 2007 in the Euro-Mediterranean Space*, IEMed/CIDOB, Barcelona.

Dernini, S. and Berry, E.M. (2015) Mediterranean Diet: From a Healthy Diet to a Sustainable Dietary Pattern, *Frontiers in Nutrition*, 2(15): 1–7.

Dernini, S., Berry, E., Serra Majem, L., La Vecchia, C., Capone, R., Medina, F. X., Aranceta, J., Belahsen, R., Burlingame,B., Calabrese, G., Corella, D., Donini, L.M., Meybeck, A., Pekcan, A.G., Piscopo, S., Yngve, A. and Trichopoulou, A. (2017) Med Diet 4.0. The Mediterranean Diet with Four Sustainable Benefits, *Public Health Nutrition*, 20(7) (May 2017): 1322–1330. doi: 10.1017/S1368980016003177.

Garine, I. de. (1993) La dieta mediterránea en el conjunto de los sistemas alimentarios. In González Turmo, I. and Romero de Solís, P. (eds) *Antropología de la alimentación: ensayos sobre la dieta mediterránea*, Junta de Andalucía/ Fundación Machado, Seville.

Garine, I. de, Teti, V., Aubaile, F., Medina, F. X., Garine, V. de, Bellio, A., Librandi, F., Scotta, D., Fidotti, M. and Cretella, E. (2005a) Recommendations Concerning the Concept of 'Mediterranean Diet'. Working paper presented to the 3rd Euro-Mediterranean Forum, *Dialogues between Civilizations and Peoples of the Mediterranean: The Food Cultures*, Rome, (np).

Garine, I. de, Teti, V., Aubaile, F., Medina, F. X., Garine, V. de, Bellio, A., , Librandi, F., Scotta, D., Fidotti, M. and Cretella, E.. (2005b) Per un'idea problematica della dieta mediterranea. Documento degli antropologi e studiosi aderenti all'International Commision for the Anthropology of Food and Nutrition (ICAF). Working paper presented to the 3rd Euro-Mediterranean Forum, *Dialogues between Civilizations and Peoples of the Mediterranean: The Food Cultures*, Rome, (np).

González Turmo, I. (1993) El Mediterráneo: dieta y estilos de vida. In González Turmo, I. and Romero de Solís, P. (eds) *Antropología de la alimentación: ensayos sobre la dieta mediterránea*, Junta de Andalucía / Fundación Machado, Seville.

González Turmo, I. (2005) Algunas notas para el análisis de las cocinas mediterráneas. In Contreras, J., Riera, A. and Medina, F. X. (eds) *Sabores del Mediterráneo: Aportaciones para promover un patrimonio alimentario común*, Institut Europeu de la Mediterrània, Barcelona.

González Turmo, I. and Medina, F. X. (2012) Défis et responsabilités suite à la déclaration de la diète méditerranéenne comme patrimoine culturel immatériel de l'humanité (Unesco), *Revue d'Ethnoécologie*, 2. Published at: https://ethnoecologie.revues.org/957. Accessed on 25 September 2016.

Gussow, J.D. (1995) Mediterranean Diets: Are They Environmentally Responsible? *American Journal of Clinical Nutrition*, 61(6): 1383–1389.

Gussow, J.D. and Clancy, K. (1986) Dietary Guidelines for Sustainability, *Journal of Nutritional Education*, 18(1): 1–5.

Herrin, M. and Gussow, J.D. (1989) Designing a Sustainable Regional Diet, *Journal of Nutritional Education*, 21(6): 270–275.

Johnston, J.J., Fanzo, J.C. and Cogill, B. (2014) Understanding Sustainable Diets: A Descriptive Analysis of the Determinants and Processes that Influence Diets and their Impact on Health, Food Security, and Environmental Sustanaibility. *Advances in Nutrition*, 5: 418–429.

Lang, T. and Barling, D. (2013) Nutrition and Sustainability: An Emerging Food Policy Discourse, *Proceedings of the Nutrition Society*, 72: 1–12.

Medina, F.X. (1996) Alimentación, dieta y comportamientos alimentarios en el contexto mediterráneo. In Medina, F. X. (ed.) *La alimentación mediterránea: Historia, cultura, nutrición*, Icaria, Barcelona.

Medina, F.X. (2009) Mediterranean Diet, Culture and Heritage: Challenges for a New Conception, *Public Health Nutrition*, 12–9 (A): 1618–1620.

Medina, F.X. (2011) Food Consumption and Civil Society: Mediterranean Diet as a Sustainable Resource for the Mediterranean Area, *Public Health Nutrition*, 14-12 (A): 2346–2349.

Medina, F. X. (2015) Assessing Sustainable Diets in the Context of Sustainable Food Systems: Socio-cultural Dimensions. In Meybeck, A., Redfern, S., Paoletti, F. and Strassner, P. (eds) *Assessing Sustainable Diets within the Sustainable Food Systems. Mediterranean Diet, Organic Food: New Challenges*, FAO, CREA and International Research Network for Food Quality and Health, Rome, pp. 159–162.

Medina, F.X., Aguilar, A. and Solé-Sedeno, J.M. (2014) Aspectos sociales y culturales sobre la obesidad. Reflexiones necesarias desde la salud pública, *Nutrición Clínica y Dietética Hospitalaria*, 34(1): 67–71.

Meybeck, A. (2015) Understanding Sustainable Diets: From Diets to Food Systems, from Personal to Global. In Meybeck, A., Redfern, S., Paoletti, F. and Strassner, P. (eds) *Assessing Sustainable Diets within the Sustainable Food Systems. Mediterranean Diet, Organic Food: New Challenges*, FAO, CREA and International Research Network for Food Quality and Health, Roma, pp. 207–214.

Serra Majem, L. (2005) La investigación epidemiológica en la Dieta Mediterránea: avances y retrocesos. In Contreras, J., Riera, A. and Medina, F. X. (eds) *Sabores del Mediterráneo: Aportaciones para promover un patrimonio alimentario común*, Institut Europeu de la Mediterrània, Barcelona, pp. 232–245.

Serra Majem, L. and Medina, F. X. (2015) The Mediterranean Diet as an Intangible and Sustainable Food Culture. In Preedy, V.R. and Watson, R.R. (eds) *The Mediterranean Diet: An Evidence-Based Approach*. Academic Press, London, pp. 37–46.

Trichopoulou, A. (2012) Diversity v. Globalization: Traditional Foods at the Epicentre, *Public Health Nutrition*, 15(6): 951–964.

Trichopoulou, A. and Lagiou, P. (1997) Healthy Traditional Mediterranean Diet: An Expression of Culture, History, and Lifestyle, *Nutrition Reviews*, 55: 383–389.

UNESCO (2010) Mediterranean Diet. Transnational Nomination: Greece, Italy, Morocco, Spain. Candidature of the Mediterranean Diet, Paris.

CHAPTER 4
CULTURES OF SUSTAINABILITY IN THE ANTHROPOCENE: UNDERSTANDING ORGANIC FOOD IN PALERMO

Giovanni Orlando

In 2013 – the year when CO_2 emissions hit the 400 ppm threshold for the first time in human history – the annual *State of the World* report (Worldwatch Institute 2013) was devoted to understanding whether achieving a sustainable society is still possible. Part two of the report, *Getting to True Sustainability*, opened with a call for the re-engineering of cultures to create a sustainable civilisation. In her contribution to it, anthropologist Melissa Leach reflects on the difficulty of achieving sustainability due to path dependencies in infrastructure (oil refineries, combine-harvesters, fridge-freezers, etc.) and social systems (corporations, governments, habits, etc.). Leach (2013: 236) writes: 'The challenge is ... to open up understanding and action around sustainability to reveal and empower alternative pathways that might currently be hidden, including those that emerge from the experiences, knowledge, and creativity of ... citizens ... in particular places'.

In this chapter I take up Leach's challenge by offering an ethnographic understanding of 'quiet sustainability' (Smith and Jehlička 2013) in the form of organic food consumption. Looking at organic foods in Palermo, Italy from a 'post-Pasteurian' perspective (Paxson 2012), I pay particular attention to those aspects concerned with the cultural perception of environmental problems and the role of science and technology in food production (Lien and Nerlich 2004). By doing so, I argue that the beliefs and feelings of people who eat organic foods allow a glimpse of the values that stem from a rejection of excess, which has several points of contact with the most recent sustainability theories of the Anthropocene (Moore 2016) and planetary boundaries (Wijkman and Rockström 2012).

Given the scope of the environmental problems at the combined earth/world-system level (Hornborg and Crumley 2006), the challenge proposed by Leach appears daunting. Studying this pressing matter with a view to including the ethnographic dimension implicit in her words is also no easy matter. And yet 'the human impact on the modern earth-system is clearly modulated by individual and group ideologies' (Galaty 2011: 19). The need to account for these competing but equally important aspects has led sustainability scholars in the social sciences to look more closely at the recursive relationship between agency, structure and culture (Spaargaren 2011). Here I focus primarily on culture as an expression of agency (Ortner 2006).

Much recent work on non-mainstream food in anthropology reflects this view of culture. This is especially the case with work on food and value (e.g. Heller 2007; Weiss 2012). What these studies explore is the difference that exists between meanings as recognised in the collective symbolic scheme generated by industrial food's political economy, and as operated on by particular groups or individuals who in various ways do not comply with the scheme as such. Initiatives such as community-supported agriculture, farm shops, vegetable-box schemes, new urban allotments, farmers' markets, solidarity purchase groups, new consumer cooperatives, Fair Trade and similar others illustrate the difference in question (Counihan and Siniscalchi 2013). The transformation of culture that these initiatives imply is key to social change towards sustainability, because it involves the renovation of those representations through which we see and act upon the world. This endeavour can profoundly benefit from the de-familiarising effect that anthropology has to offer. A trend in this direction appears to be developing among anthropologists, though admittedly with quite diverse premises and intentions (e.g. Isenhour 2010; Miller 2012).

The chapter is organised as follows: in the first section I set out the broad context in which Palermitan organic consumers' opinions are formed; the second section deals with their lay notions of food risk; the third one discusses more specifically the idea of accumulating pollution; the fourth section relates informants' opinions to the latest findings in sustainability theory; and the fifth examines what factors explain this relation. The conclusions follow.

Food in a Post-Pasteurian World

According to a recent study (Benbrook 2016), glyphosate, the active ingredient in the famous Monsanto herbicide Roundup®, has become the most widely used agricultural chemical of all time. Since its introduction in 1974, 8.6 million tons of it have been sprayed worldwide; in 2014 alone, enough glyphosate was sprayed to apply almost half a pound of it on every cultivated acre of land. In 2016, the European Parliament had to vote on whether to re-licence the substance. As I write, the licence has been extended by a mere eighteen months, due to the inability of the Parliament to reach an agreement on the issue. To raise awareness in favour of its banning, forty-eight Members of the European Parliament gave urine samples to see if they contained traces of the substance. The results indicated an average concentration of 1.7 micrograms/litre, which is seventeen times higher than the European drinking water norm.

The test was not intended to prove a link to human pathology. Nevertheless, the controversy it generated was considerable, especially given the uncertainty of more official sources of information. The World Health Organisation, for example, considers glyphosate 'probably carcinogenic' (Guyton et al. 2015), while the European Food Safety Authority thinks it is 'unlikely to pose a public health risk' (Neslen 2015). In 2014, official tests carried out by the expert committee on pesticide residues for the United Kingdom government found traces of glyphosate in two-thirds of the bread loaves sold in supermarkets (Carrington 2014), but concluded that they did not have a detrimental effect on health. Members of the UK's Pesticide Action Network disagreed, and suggested that if people wanted to avoid ingesting pesticides then they should eat organic food. Their advice followed a major study on the differences between organic and conventional foods (Bara ski et al. 2014), which concluded that pesticides were found four times more often on conventional fruit and vegetables.

This glyphosate controversy is indicative of the kind of world in which organic food acquires significance. The scientific debate regarding the potential threat from invisible hazards, the political dispute over market regulation and the recourse to technology to substantiate an ethical position (the precautionary principle) are all defining characteristics of a post-Pasteurian world (Paxson 2012: 161). In such a world, society is no longer able to maintain the kind of dominance over nature symbolised by the work of Pasteur (Latour 1988), a dominance that had generated all the benefits of modernity. Paradoxically, this incapacity is caused by the negative impacts on nature that society itself has generated, which now come back to plague it in the form of all sorts of risks – for example, food risk (Paxson 2012: 184). The premises that underpinned industrialisation are no longer valid in this age, and life has become risky due to the breakdown of the boundaries between nature and culture. I was often reminded of these ideas during conversations I had with Palermitans in the field.[1]

Lay Notions of Food Risk

Brigida was a 48-year-old science teacher whom I met while working in one of Palermo's health food stores. This store was run by a small workers' cooperative, its modest size placing it in the category of neighbourhood shops.

The goods on sale included fresh and processed organic foods (fruit and vegetables, pasta, breakfast food, snacks, etc.), Fair Trade foods, ecological cleaning products (e.g. washing-up liquid) and personal care items (soaps, shower gels, etc.). After a few months of acquaintance, I asked Brigida if I could visit her at home to discuss her shopping habits and her reasons for eating organic foods.

Sitting at her kitchen table, Brigida began elaborating on her food purchases and, as soon became clear, on much more than that. She started by saying that she was interested in food education both as a result of her school job and as an agricultural science graduate (though she made no use of her formal, scientific knowledge of food production at the school). Then she said: 'So one day I realised that a whole series of foods had become inedible. What triggered my desire to buy organic was the idea that by shopping in supermarkets there was no chance of having a healthy diet'. Brigida believed that supermarket food was full of what she defined as 'substances that cause health problems'. Explaining this belief, she mentioned mostly short-term but recurrent problems, like colitis, food poisoning, rashes and allergies. But she also talked of people who had died of cancer when they were just fifty years old because they could not eat properly. For these reasons, her organic purchases were inspired by 'a philosophy of risk reduction'. She said: 'I try to lower the dose of poisons, because I'm convinced that half of all diseases originate in food. I'm not saying it – the studies say it. So, if you control the quality of food, you lower the risk of a whole series of problems'.

Brigida's concerns echoed those of most Palermitans I met who ate organic food. This group of people considered conventional agriculture responsible for the unsafe state of almost all the food found in mainstream retailers. Through the sickness it was believed to cause, so-called dangerous conventional food forced these Palermitans to reflect on their shopping choices and their diets. Risk and reflexivity were thus firmly tied together, because informants constantly needed to think about how to avoid being harmed. However, this self-reflexivity did not entail the capacity to perceive risk entirely by their own means. Compared to the environmental dangers of the past, which 'assaulted the nose or the eyes and were thus perceptible to the senses', the 'risks of civilisation today escape perception and are localised in the sphere of physical and chemical formulas (e.g. toxins in foodstuffs)' (Beck 1992: 21).

The invisible aspect of food risk explains informants' reference to scientific knowledge. Brigida, for example, referred to the 'studies' that prove the link between food quality and illness. A 29-year-old lawyer, Simona, mentioned Hippocrates as the first 'scientist' who had made the connection between food and health more than two millennia ago. While a 53-year-old psychiatrist, Annamaria, said that the label 'organic' meant 'what's written in books: the absolute absence of pesticides, of genetically modified organisms (GMOs) and other horrible stuff'.

Food and Pollution

In some form or another, almost all informants repeated Brigida's comments on 'poisons' or Annamaria's on 'horrible stuff'. These Palermitans usually referred to harmful substances through negative reasoning, by citing their *absence* in organic food. Their comments were thus reminiscent of what Pratt (2008: 62) has written about alternative food imaginaries: 'In these matters we get the sense of a self-confirming semantic field ... that we will only get a handle on ... by spelling out what it is defined against'. The specific substances mentioned varied, but usually belonged to the same group: pesticides, GMOs, hormones, antibiotics, additives, colorants. The following example is illustrative of this.

Lorenzo was a 36-year-old salesman who worked for a company that installed solar panels, whom I met through a solidarity purchase group that delivered a weekly box of organic produce.[2] He told me he was against the use of GMOs because there were still not enough studies on the potential long-term effects of their consumption by humans. At one point during our conversation, Lorenzo asked me rhetorically whether I would eat something knowing that it could be poisonous. When I replied negatively, he sealed his point emphatically by saying: 'But what's the difference if you get cancer in thirty years' time?'

Lorenzo's case is notable for at least three reasons. First, he referred to three topics that were also prominent in Brigida's line of reasoning, using exactly the same words: studies, poisons and cancer. Secondly, his implicit suggestion that one should avoid an action in the present (eating conventional food) because of a danger ensuing in the future (developing cancer), is a typical example of the kind of thought process that guides everyday life in a post-Pasteurian world: while the present/past determined the future during early modernity, with the onset of late modernity it is the future that determines the present.

The temporally indeterminable effects of current risks cause this logic inversion, which leads us to the third reason of interest in Lorenzo's argument: the systemic and accumulating nature of risks, particularly 'pollution'.[3] His claim that I might get cancer in thirty years' time clearly depicted the effects of harmful substances as cumulative. For many informants, the fear of accumulating pollution resulted from the daily, repetitive act of eating, which translated into daily exposure. This argument was also suggested to me by the 41-year-old social worker, Martina, when she offered this reply to those who criticise organic foods by saying that pollution is unavoidable because it is now endemic: 'Well, it's always less [pollution] than what you consume. Then of course, you haven't grown them in Heaven!'

These Palermitans were concerned about food as a result of a more widespread concern for pollution, food being a privileged channel through which they could absorb it. This group of people feared food because it contained too much 'horrible stuff', to borrow Annamaria's expression. This 'stuff' accumulated constantly, making not just the present risky, but even the future. A concern for excess, for the exceeding of limits, was therefore a common denominator among them. Recently there have been new calls in sustainability studies to pay attention to the problem of excess. One of these is the idea of planetary boundaries, which I explain in the following section.

Planetary Boundaries and the Anthropocene

The planetary boundaries approach is a set of tools for assessing the current prospects of sustainability. The planet's state is self-regulated by a number of biophysical processes. When key variables in them exceed certain thresholds due to global or local-cumulative impacts, non-linear transitions are triggered that are by and large uncontrollable and unpredictable. Planetary boundaries are therefore the safe 'operating spaces' inside which human activity should remain in order to preserve a suitable habitat. 'Boundaries … are human determined values of the control variable set at a "safe" distance from a dangerous level' (Rockström et al. 2009: 3).

The unfettered growth of human economic activities is causing worry that more pressure on the planet could disrupt key biophysical structures and set in motion irrevocable changes that would be devastating for our society (Rockström et al. 2009: 2). This catastrophic change could shift the Earth away from the environmental patterns of the Holocene, the geologic era under which all major elements of civilisation have been built, towards a new unpredictable era that will pose enormous challenges of adaptation. Many authors believe that we have already entered this new era, which has been named the Anthropocene to underscore humanity's impact on the planet as a global force (Hamilton, Bonneuil and Gemenne 2015).

Scientists have identified nine planetary boundaries (Steffen et al. 2015), three of which are directly linked to the industrial food system, particularly with regard to pollution, and are therefore of relevance to Palermitans' concerns.[4] The first one is greenhouse gases, which are connected to the agri-food system via fertiliser and pesticide use, mechanisation, transportation, packaging, refrigeration, and such like. The second one is the global cycle of nitrogen and phosphorous, the vast majority of which goes to the agricultural sector as fertilisers. The third boundary is represented by so-called novel entities (previously termed 'chemical pollution', see Rockström et al. 2009: 18), which include heavy metals and organic compounds of human origin. Greenhouse gases, nitrogen and phosphorous have already exceeded their safe limits, moving humanity deeper into the Anthropocene.

Specific planetary boundaries, then, are clearly linked to the world of food and to Palermitans' apprehensions about it. Palermitans' concern about excessive pollution, manifested through the medium of organic foods, resonates closely with this strand of ecological thinking. Anthropologically, it can be considered part of a culture of sustainability. And yet food anxieties are often criticised for being egoistic and individualistic. Most consumer research, for example, sees having a healthy diet as the primary factor pushing consumers to buy organic food, with environmental concerns being only secondary (e.g. Hughner et al. 2007).

Consumer research has also influenced understandings of the organic phenomenon in the social sciences, where sharp divisions between personal and altruistic motives are often found. In an early contribution to the study of organic foods, for example, James (1993: 213) proposed a continuum of reasons that spanned the poles of environmentalism and lifestyle, concluding: 'The rise of an organic food market cannot necessarily be seen as an indication of the integration of environmentalist principles'. While this may be true, environmental*ism* and having concerns for (some aspect of) the environment are two different matters. Perhaps organic consumption is not a form of environmentalism, but it could be considered a form of 'quiet sustainability'.

Cultures of (Quiet) Sustainability

The concept of quiet sustainability (Smith et al. 2015) refers to everyday practices that can have a beneficial environmental role, but that are not represented by those who perform them as relating directly to sustainability. Looking at the Palermitan material through this lens allows one to find common elements between the ideas of organic consumers and the theory of planetary boundaries. I would suggest three points of contact.

First, there is the idea that the cumulative effect of manufactured risks swamps our capacity to deal with them. Effectively, this is the result of the biosphere's (in)capacity to cope, which humans cannot replace. Risk both originates in, and leads to, excess. Second, the ubiquity and permanence of environmental damage may indeed still cause catastrophic consequences. This is not catastrophism for its own sake – there is real evidence supporting such claims (Sale 2011). If anything, the problem is negationism. Finally, there is the idea that human economic activities create not just goods but also 'bads' – the negative environmental consequences of commodity production – which lead to the exceeding of global biosphere limits.

What accounts for these points of contact? In a post-Pasteurian world, the complex dynamic between the media, activists and scientists plays a key role in giving rise to the perception of environmental pollution as ubiquitous. The same dynamic was behind the birth of the modern organic movement, which explains, at least in part, the common elements between the two social phenomena. In a sense, the Palermitans I spoke to were conveying in their own words the concerns on which the organic food movement itself was founded. This shared discourse can be explained by the fact that, while consuming organic food, Palermitans came into contact with bits and pieces of relevant information.

Working in health food stores, I was able to witness this process first-hand. All the items on sale in the stores conveyed risk themes in one form or another. Organic foods reminded shoppers of their official certification, which guaranteed a lack of chemical residues, GMOs, and artificial ingredients more generally. Similar statements were made on Fair Trade foods, many of which were also certified organic. Eco-friendly cleaning products indicated they did not damage rivers or the ocean, thanks to the use of food- or plant-based components. Personal care items, many of which included organic ingredients, reassured the buyer of their non-irritant nature, thanks to tests that had revealed no traces of harmful substances. I saw similar evidence when visiting people's homes, in the form of books, newspaper articles, magazines and emails. Ultimately, all this kind of information draws loosely on an ecological critique that has many elements in common with the theories discussed above.

The common ground between the organic consumer ideology and the environmental discourses in question can also be explained historically. Organic food rose to prominence in the 1970s amidst new awareness about environmental threats. One of the most dramatic illustrations of this period was the reaction to Rachel Carson's book *Silent Spring*, which made public the impact of pesticides on biodiversity, eventually leading to the banning of DDT and other insecticides. Another source of concern at the time was one of the 'blue baby' syndrome causes, methaemoglobinaemia, due to elevated nitrate levels in drinking water, generally caused by high applications of fertilisers to soil in which edible plants had been grown. In 1972, the United Nations convened the first global environmental summit in Stockholm amid international controversies such as the pollution of Japan's Minamata Bay with mercury, the US military's use of defoliant chemicals in Vietnam, and the acid rains resulting from sulphur and nitrogen emissions. This social context prompted a growing interest in foods that were perceived to be cleaner, like organics. It would seem that, more than forty years on, we are still living through the same social history.

Conclusions

My aim in this chapter has been to advance an understanding of global environmental change through the life experiences of a particular group of people in one locale. To do so, I have looked at organic food as a lens through which to capture elements of a possible culture of sustainability, based – largely implicitly or 'quietly' – on anxieties about pollution and the existence of limits to the human modification of nature. Drawing on a post-Pasteurian perspective and the planetary boundaries approach, I have shown how the apprehension about excessive pollution among the Palermitans I met reflects current ecological thinking, due to the shared origins of the organic movement and (parts of) the environmental one. While this chapter discusses a case study of the anthropology of food, I also sought to go beyond the anthropological analysis of food value, motivated by what Hornborg (2011: 8) describes as 'the important but difficult ambition to bridge the divide between local particularities of experience ... and universalising understandings of global socio-ecological processes'. The latest work on alternative food movements shows that these can provide an apt lens for this kind of analysis, albeit a rarely used one (Nonini 2013). There is a need within the anthropology of food to make a collaborative effort to demystify ideologies of growth, development and consumption, as it is their material repercussions that are causing the biosphere's degradation.

Acknowledgements

The research discussed in this chapter received financial support from the UK Economic and Social Research Council (Grant PTA-030-2006-00260) and from the Royal Anthropological Institute (Emslie Horniman Anthropological Scholarship Fund).

Giovanni Orlando
University of Turin, Italy

Notes

1. This chapter is based on extensive participant observation in Palermo and its rural province, lasting a total of fifteen months. Participant observation took the form of voluntary work in organic food stores and on organic farms, household research (observation of food shopping trips and meal preparation, guided tours of food-related house spaces, consumption inventories), and attendance of group meetings and public events. In addition to participant observation, thirty-three tape-recorded, semi-structured interviews were carried out, mostly with women aged between 29 and 53, often married and with children.
2. Solidarity purchase groups are a distinctively Italian version of short food supply chains (see Grasseni 2013).
3. I am using the word pollution in a broad sense here, both as 'matter out of place' (Douglas 1966) and in a more objectivist sense, as actual toxic materials.
4. A fourth boundary, biosphere integrity, is also directly relevant to the world of food. However, as it is not in itself an issue of pollution, I will not dwell on it.

References

Barański M., Średnicka-Tober, D., Volakakis, N., Seal, C., Sanderson, R., Stewart, G.B., Benbrook, C., Biavati, B., Markellou, E., Giotis, C., Gromadzka-Ostrowska, J., Rembiałkowska, E., Skwarło-Sońta, K., Tahvonen, R., Janovská, D., Niggli, U., Nicot, P. and Leifert, C. (2014) Higher Antioxidant and Lower Cadmium Concentrations and Lower Incidence of Pesticide Residues in Organically Grown Crops: A Systematic Literature Review and Meta-analyses, *British Journal of Nutrition*, 112(5): 794–811.

Beck, U. (1992) *Risk Society: Towards a New Modernity*, Sage, London.

Benbrook, C.M. (2016) Trends in Glyphosate Herbicide Use in the United States and Globally, *Environmental Sciences Europe*, 28(3): 1–15.

Carrington, D. (2014) Over 60% of Breads Sold in the UK Contain Pesticide Residues, Tests Show. Published at: http://www.theguardian.com/environment/2014/jul/17/pesticide-residue-breads-uk-crops. Accessed on 8 October 2016.

Counihan, C. and Siniscalchi, V. (eds) (2013) *Food Activism: Agency, Democracy, Economy*, Bloomsbury, London.

Douglas, M. (1966) *Purity and Danger: An Analysis of Concepts of Pollution and Taboo*, Routledge, London.

Galaty, M.L. (2011) World-Systems Analysis and Anthropology: A New Détente? *Reviews in Anthropology*, 40(1): 3–26.

Grasseni, C. (2013) *Beyond Alternative Food Networks: Italy's Solidarity Purchase Groups*, Bloomsbury, London.

Guyton, K.Z., Loomis, D., Grosse, Y., El Ghissassi, F., Benbrahim-Tallaa, L., Guha, N., Scoccianti, C., Mattock, H. and Straif, K. (2015) Carcinogenicity of Tetrachlorvinphos, Parathion, Malathion, Diazinon, and Glyphosate, *The Lancet Oncology*, 16(5): 490–491.

Hamilton, C., Bonneuil, C. and Gemenne, F. (eds) (2015) *The Anthropocene and the Global Environmental Crisis*, Routledge, London.

Heller, C. (2007) Techne versus Technoscience: Divergent (and Ambiguous) Notions of Food 'Quality' in the French Debate over GM crops, *American Anthropologist*, 109(4): 603–615.

Hornborg, A. (2011) *Global Ecology and Unequal Exchange: Fetishism in a Zero-sum World*, Routledge, London.

Hornborg, A. and Crumley, C.C. (eds) (2006) *The World System and the Earth System: Global Socioenvironmental Change and Sustainability since the Neolithic*, Left Coast Press, Walnut Creek, CA.

Hughner, R.S., McDonagh, P., Prothero, A., Shultz II, C.J. and Stanton, J. (2007) Who Are Organic Food Consumers? A Compilation and Review of Why People Purchase Organic Food, *Journal of Consumer Behaviour*, 6: 94–110.

Isenhour, C. (2010) Building Sustainable Societies: A Swedish Case Study on the Limits of Reflexive Modernization, *American Ethnologist*, 37(3): 511–525.

James, A. (1993) Eating Green(s): Discourses of Organic Food. In Milton, K. (ed.) *Environmentalism: The View from Anthropology*, Routledge, London, pp. 203–215.

Latour, B. (1988) *The Pasteurization of France*, translated by Alan Sheridan and John Law, Harvard University Press, Cambridge, MA.

Leach, M. (2013) Pathways to Sustainability: Building Political Strategies. In Worldwatch Institute *State of the World 2013: Is Sustainability Still Possible?* Island Press, Washington, DC, pp. 234–243.

Lien, M.E. and Nerlich, B. (eds) (2004) *The Politics of Food*, Berg, Oxford.

Miller, D. (2012) *Consumption and Its Consequences*, Polity, Cambridge.

Moore, A. (2016) Anthropocene Anthropology: Reconceptualizing Contemporary Global Change, *Journal of the Royal Anthropological Institute*, 22(1): 27–46.

Neslen, A. (2015) EU Watchdog Opens Door to New Licence for Controversial Weedkiller. Published at: http://www.theguardian.com/environment/2015/nov/12/eu-watchdog-approves-new-license-for-controversial-weedkiller. Accessed on 8 October 2016.

Nonini, D.M. (2013) The Local-Food Movement and the Anthropology of Global Systems, *American Ethnologist*, 40(2): 267–275.

Ortner, S.B. (2006) *Anthropology and Social Theory: Culture, Power, and the Acting Subject*, Duke University Press, Durham, NC.

Paxson, H. (2012) *The Life of Cheese: Crafting Food and Value in America*, University of California Press, Berkeley.

Pratt, J. (2008) Food Values: The Local and the Authentic, *Research in Economic Anthropology*, 28: 53–70.

Rockström, J., Steffen, W., Noone, K., Persson, Å., Chapin III, F.S., Lambin, E., Lenton, T.M., Scheffer, M., Folke, C., Schellnhuber, H., Nykvist, B., de Wit, C.A., Hughes, T., van der Leeuw, S., Rodhe, H., Sörlin, S., Snyder, P.K., Costanza, R., Svedin, U., Falkenmark, M., Karlberg, L., Corell, R.W., Fabry, V.J., Hansen, J., Walker, B., Liverman, D., Richardson, K., Crutzen, P. and Foley, J. (2009) Planetary Boundaries: Exploring the Safe operating Space for Humanity, *Ecology and Society*, 14(2): 32.

Sale, P.F. (2011) *Our Dying Planet: An Ecologist View of the Crisis We Face*, University of California Press, Berkeley.

Smith, J. and Jehlička, P. (2013) Quiet Sustainability: Fertile Lessons from Europe's Productive Gardeners, *Journal of Rural Studies*, 32: 148–157.

Smith, J., Kostelecký, T. and Jehlička, P. (2015) Quietly Does It: Questioning Assumptions about Class, Sustainability and Consumption, *Geoforum*, 67: 223–232.

Spaargaren, G. (2011) Theories of Practices: Agency, Technology and Culture: Exploring the Relevance of Practice Theories for the Governance of Sustainable Consumption Practices in the New World-Order, *Global Environmental Change*, 21(3): 813–822.

Steffen, W., Richardson, K., Rockström, J., Cornell, S.E., Fetzer, I., Bennett, E.M., Biggs, R., Carpenter, S.R., de Vries, W., de Wit, C.A., Folke, C., Gerten, D., Heinke, J., Mace, G.M., Persson, L.M., Ramanathan, V., Reyers, B. and Sörlin, S. (2015) Planetary Boundaries: Guiding Human Development on a Changing Planet, *Science*, 347(6223): 1–16.

Weiss, B. (2012) Configuring the Authentic Value of Food: Farm-to-Fork, Snout-to-Tail, and Local Food Movements, *American Ethnologist*, 39(3): 614–626.

Wijkman, A. and Rockström, J. (2012) *Bankrupting Nature: Denying Our Planetary Boundaries*, Earthscan, London.

Worldwatch Institute (2013) *State of the World 2013: Is Sustainability Still Possible?* Island Press, Washington, DC.

CHAPTER 5
WILD PHYTOGENETIC RESOURCES FOR FOOD IN THE BARRANCA DEL RÍO SANTIAGO, MEXICO: A FIRST APPROACH TO SUSTAINABILITY

....

Martín Tena Meza, Rafael M. Navarro-Cerrillo, Ricardo Ávila Palafox and Raymundo Villavicencio García

Introduction

The manner in which food production is carried out can compromise the capacity of land to provide food for all those on the planet, and the ambition in the twenty-first century for food security through the production of sustainable foods requires, among other things, environmental sustainability in terms of biodiversity. For this reason, to conserve and organise biodiversity becomes a priority activity, as understood in the International Treaty on Plant Genetic Resources for Food and Agriculture adopted by the Thirty-First Session of the Conference of the Food and Agriculture Organization of the United Nations on 3 November 2001[1] and the Second Global Plan of Action for Plant Genetic Resources for Food and Agriculture.[2]

Bearing this in mind, wild plants can be put to a number of uses, among which the following stand out:

- their nutritional use
- the opportunity to create new crops
- their potential to be included into sustainable development programmes.

Within Mexican tropical vegetation, deciduous forest is the most extensive. It holds third place in national plants diversity (Flores-Vilella and Gerez 1994: 9). In spite of its environmental relevance, it suffers a high deforestation rate and nowadays only 30 percent of the original surface remains in a relatively good state of preservation (Trejo 2005: 1). This paper presents an outline of the wild and sociocultural conditions of the phytogenetic alimentary resources of the prevailing dry deciduous forest in Barranca del Río Santiago (BRS), a unique canyon in western Mexico, in which an area of approximately 723 square kilometres was selected for studying.

Our chapter will present some preliminary findings about botanical data from our work in the canyon, supplemented by interviews with local people about their food habits. The objective of our chapter is to introduce the first part of a database on phytogenetic resources for food in BRS, created to generate ethnobotanical and ethnoecological records on the canyon's flora. It also shows some worrying aspects about the flora's leveraging.[3] Considering that the salient features of the dry deciduous forest are trees and bushes, the first segment of the database only takes into account trees and bushes with one or several structures that are used, now and then, by the inhabitants of the canyon for alimentary purposes. Herbaceous plants will be addressed in another document. Evidence concerning the presence and use of these resources will prove useful in designing alimentary plans, sustainable farming and environmental programmes for the sustainable development of the villages surrounding the canyon, and to support handling the agave landscape as a cultural heritage of humanity and the treatment of the protected natural areas by different government areas within the study area.

The Area and Its Botanical Resources

The BRS is located in western Mexico, on the northern limits of what cultural anthropologists call the Mesoamerican cultural area (Kirchhoff 1967: 1–7). The zone has great significance because it represents an area of strong interaction between Holarctic and Neotropical kingdoms, which constitutes a biological mix and is a corridor between the temperate ecosystems of central Jalisco and the tropical environments of the Pacific coast (Challenger 1998: 283) (Figure 5.1). This is important because the zone's physiographic constitution, the degree of cultural development achieved by the pre-Colombian peoples, who inhabited it, and its rich phytogenetic resources combined to produce a space where a broad range of plants were domesticated (Vargas-Ponce et al. 2007:363, 373; Zizumbo-Villareal et al. 2014:68). Moreover, it was a corridor through which human groups moved, diffusing ideas and alimentary habits.

The predominant vegetation in the BRS is a variant of the 'dry jungle system', or 'deciduous tropical forest' (DTF), and is comprised of low-growing forest vegetable communities that prosper on hillsides in warm climatic conditions, with a pluvial regimen that lasts for six months or less, in a cycle divided into two main periods: the dry season and the rainy season (Rzedowsky and Calderon 2013: 2). This type of vegetation is distinguished by the numerous ecoregions, habitats and ecosystems it contains, and it offers a beneficial

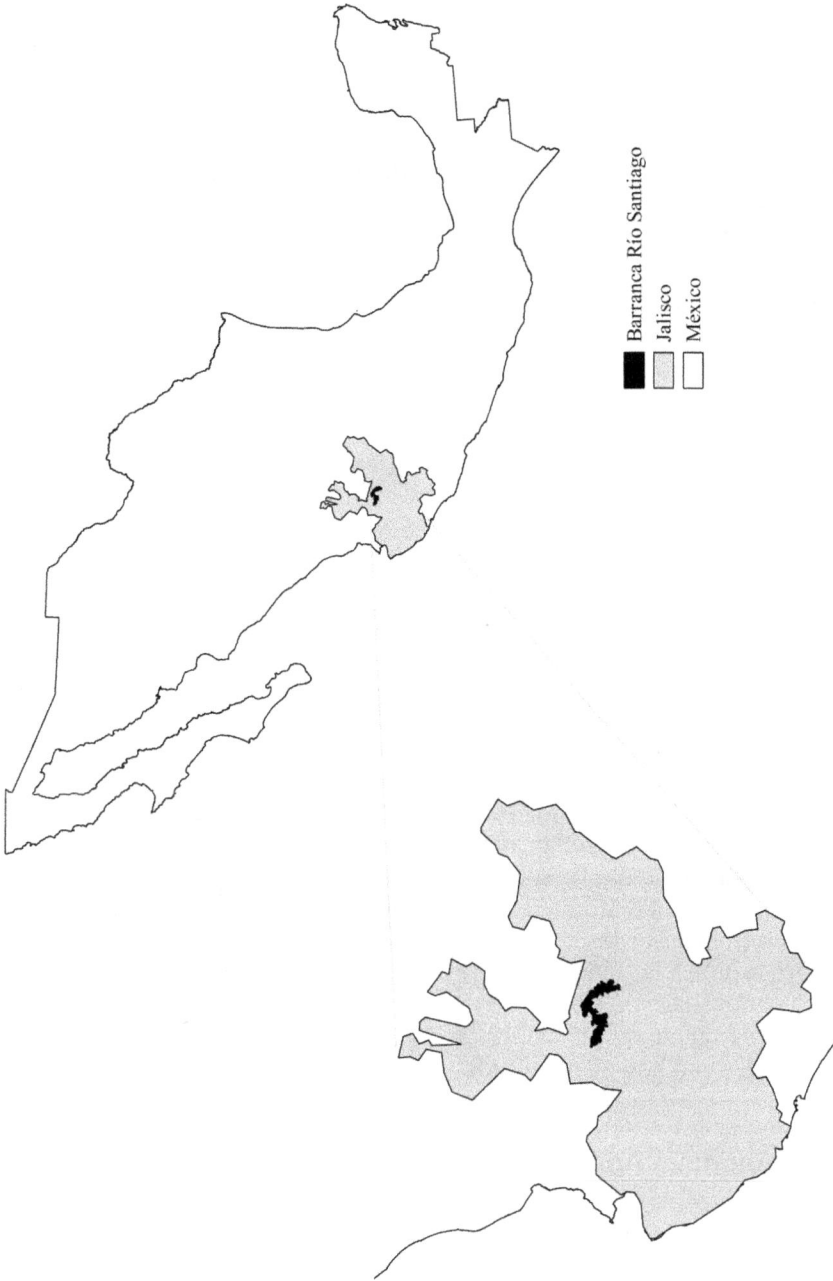

Figure 5.1 Map of the Barranca del Río Santiago study area. Figure created by the author.

environment to the many people who exploit its bounty. In this type of environment, inhabitants have varied uses for most of the plants – in some cases, for over 60 percent of them (Dorado 2000, cited by Soto 2010: 285).

The botanical explorations of the region for scientific purposes date back to 1791 with Sessé and Mociño's *Real Expedición Botánica a Nueva España* [Royal Botanical Expedition in New Spain] (McVaugh 1972: 208), but it was not until the mid-nineteenth century when those articles about Mexican foliage first appear in the literature; some of these articles relate it to altitude, climate and/or other environmental factors (Rezedowski 2006: 13). For over two hundred years, numerous botanists of national and foreign origin have taken an interest in Jalisco's flora; but the records are far from complete and our knowledge of the region's flora remains inadequate. To date, botanical studies have been published on only 23 of the estimated 166 families of dicotyledonous plants that exist in the State of Jalisco, which, according to the *Informe Nacional sobre el Estado de los Recursos Fitogenéticos para la Agricultura y la Alimentación* [National Report on the State of Plant Genetic Resources for Agriculture and Food], is justifiably recognised for its rich biodiversity, even though the floristic and ethnobotanical studies, required to understand this in terms of conservation and sustainable exploitation, have not yet been conducted (Molina and Córdova 2006:17, 57, 87). Therefore, our study is aimed at work on this.

Recent botanical studies carried out at different sites of the BRS, divided in sections near Guadalajara, have helped to increase our knowledge of the zone's flora. For example, in just one of those sections, researchers recorded 869 species, 47 of which are endemic to this area and unique in Mexico (Universidad de Guadalajara 2009). Of these, six are classified as 'protected species' because their existence is deemed to be 'threatened'. Unfortunately, at present, there are only two guides to arboreal species and a few isolated botanical studies that together represent only a modest contribution towards efforts to systematise our knowledge of the area (López et al. 2011; Sahagún-Godínez et al. 2014). Although Mexico has signed several international protocols for biological conservation, initiatives undertaken in this regard in the BRS are insufficient. In summary, it is necessary, and urgent, to increase not only the knowledge of in situ floristic species, but also, through ethnobotanical studies, to increase our understanding of local human attitudes towards these species and how such attitudes may be changing.

Methodological Approach

In light of the lack of a catalogue of vascular plants or any inventory of edible plants in the specific area of study, a database has been constructed to register pertinent species reported in botanical studies performed in situ, and in neighbouring areas with similar features (Secretaría de Medio Ambiente y

Desarrollo Territorial 2016: 313–362). Furthermore, to assess the presence of these plants and the villagers' interaction with them, more than a hundred ethnographic forays into the canyon (72,300 Ha) were made with an ethnobotanical approach, focusing on the human settlements in accessible areas relevant for the purposes of our investigation. More than seventy inhabitants were interviewed and grouped through focused questionnaires and open-ended, but directed, interviews, most of which involved women and elderly people. The main objective of those interviews was to determine which foods people in the zone had eaten in the past, and which ones are still included in their current diet.

Data from both assignments were reviewed and checked against literature from other ethnobotanical investigations, both within and outside of Mexico, in order to verify the presence of each species within the study area and the reported use for alimentary purposes of every one of the species integrated into the final database. These data are complemented by the results of field surveys, our own botanical collections and information from numerous inhabitants of the area gathered through the questionnaires, in order to review the extent to which this was based on local flora and, so, the sustainability of these species.

Results

Following analysis of the 1,218 entries in the original database, every single one of the species was examined considering: (1) the ethnobotanical knowledge about the taxonomic family to which it belongs; (2) its morphologic characteristics; (3) any previous physical-chemical analysis; and (4) the background of its alimentary use within the study area or any other region.

In a second review, species were eliminated from the database if their presence in the study area (BRS) could not be confirmed by the interviewees, nor recognised and collected during the ethnobotanical exploration, nor listed in any herbarium or in any specific flora listing. Eventually only registered species present within the study area and with confirmed evidence of present-day uses in or outside of the study area were included in the database.

By following this methodology, the analysis of the 1,218 entries has led to the identification of 172 wild edible species (herbaceous and woody plants)[4] distributed in 50 botanical families, the most prominent of which are *fabaceae*, *cactaceae* and *solanacea*. It is important to note that this number is far from insignificant and that these 172 represent 11 percent of the approximately 1,500 edible vegetable species recorded in Mexico (Casas 2010: viii). Of the 172 species, 96 are trees or bushes with fruits, foliage or roots or edible starches. Half of these (48) are listed as edible in the two existing flora guides on the BRS (López et al. 2011; Sahagún-Godínez et al. 2014). The other half are found within the inventories of the investigative exploration (Table 5.1).

Table 5.1 Edible trees and shrubs in the Barranca del Río Santiago, Jalisco, Mexico.

Family/*Species*	Shape	Origin
Anacardiaceae		
Mangifera indica L.	Arborescent	I
Rhus jaliscana Standl.	Shrubby	N
Spondias mombin L.	Arborescent	N
Spondias purpurea L.	Arborescent	N
Annonaceae		
Annona cherimola Mill.	Arborescent	N
Annona longiflora S. Watson	Arborescent	N
Annona reticulata L.	Arborescent	N
Annonna muricata L.	Arborescent	N
Apocynaceae		
Stemmadenia donnell-smithii (Rose) Woodson	Arborescent	N
Stemmadenia tomentosa Greenm.	Arborescent	N
Thevetia ovata (Cav.) A. DC.	Arborescent	N
Thevetia thevetioides (Kunth) K. Schum.	Arborescent	N
Bignoniaceae		
Crescentia alata Kunth	Arborescent	N
Bombacaceae		
Ceiba aesculifolia (Kunth) Britten & Baker f.	Arborescent	N
Pseudobombax palmeri (S. Watson) Dugand	Arborescent	N
Boraginaceae		
Ehretia latifolia Loisel	Arborescent	N
Cactaceae		
Cephalocereus alensis (F.A.C. Weber) Britton & Rose	Arborescent	N
Isolatocereus dumortieri (Scheidw.) Backeb.	Arborescent	N
Opuntia atropes Rose	Shrubby	N
Opuntia fuliginosa Griffiths	Shrubby	N
Opuntia jaliscana Bravo	Shrubby	N
Opuntia pubescens J.C. Wendl. ex Pfeiff.	Shrubby	N
Opuntia undulata Griffiths	Shrubby	N
Pachycereus pecten-aboriginum (Engelm. ex S. Watson) Britton & Rose	Arborescent	N
Pilosocereus alensis (F.A.C. Weber) Byles & G.D. Rowley	Arborescent	N
Stenocereus dumortieri (Scheidw.) Buxb.	Arborescent	N
Stenocereus queretaroensis (F.A.C. Weber) Buxb.	Arborescent	N

Cannabaceae
   Aphananthe monoica (Hemsl.) J.-F. Leroy   Arborescent   N
   Celtis caudata Planch.   Arborescent   N
   Celtis iguanaea (Jacq.) Sarg.   Arborescent   N

Species	Habit	
Cannabaceae		
Aphananthe monoica (Hemsl.) J.-F. Leroy	Arborescent	N
Celtis caudata Planch.	Arborescent	N
Celtis iguanaea (Jacq.) Sarg.	Arborescent	N
Capparaceae		
Crateva palmeri Rose	Arborescent	N
Caricaceae		
Carica papaya L.	Arborescent	N
Jacaratia mexicana A. DC.	Arborescent	N
Ebenaceae		
Diospyros digyna Jacq.	Arborescent	N
Euphorbiaceae		
Dalembertia populifolia Baill.	Shrubby	N
Manihot aesculifolia (Kunth) Pohl	Shrubby	N
Manihot caudata Greenm.	Shrubby	N
Manihot rhomboidea Müll. Arg.	Shrubby	N
Fabaceae		
Acacia acatlensis Benth.	Arborescent	N
Acacia cochliacantha Humb. & Bonpl. ex Willd.	Shrubby	N
Acacia coulteri Benth.	Arborescent	N
Acacia farnesiana L. Willd.	Arborescent	N
Enterolobium cyclocarpum (Jacq.) Griseb.	Arborescent	N
Erythrina flabelliformis Kearney	Arborescent	N
Erythrina montana Rose & Standl.	Arborescent	N
Inga vera Willd.	Arborescent	N
Leucaena esculenta (Moc. & Sessé ex DC.) Benth.	Arborescent	N
Leucaena leucocephala Lam.	Arborescent	N
Pithecellobium dulce (Roxb.) Benth.	Arborescent	N
Prosopis laevigata (Humb. & Bonpl. ex Willd) M.C.Johnst.	Arborescent	N
Tamarindus indica L.	Arborescent	I
Fagaceae		
Quercus rugosa Née	Arborescent	N
Juglandaceae		
Juglans regia L.	Arborescent	I
Lamiaceae		
Hyptis albida Kunth	Shrubby	N
Vitex mollis Kunth	Arborescent	N
Vitex pyramidata B.L. Rob. ex Pringle	Arborescent	N

Lauraceae
Persea americana Mill.   Arborescent   N
Malpighiaceae
Malpighia mexicana A. Juss   Shrubby   N
Malvaceae
Gossypium aridum (Rose & Standl.) Skovst   Shrubby   N
Guazuma ulmifolia Lam.   Arborescent   N
Hibiscus phoeniceus Jacq.   Shrubby   N
Malvaviscus arboreus Cav.   Shrubby   N
Mirtaceae
Psidium sartorianum (O. Berg) Nied.   Arborescent   N
Psidiuma guajaba L.   Arborescent   N
Moraceae
Brosimum alicastrum Sw.   Arborescent   N
Ficus goldmanii Standl.   Arborescent   N
Ficus insipida Willd.   Arborescent   N
Ficus pertusa L.   Arborescent   N
Ficus pringlei S. Watson   Arborescent   N
Morus celtidifolia Kunth   Arborescent   N
Olacaceae
Ximenia parviflora Benth.   Shrubby   N
Oleaceae
Forestiera tomentosa S. Watson   Arborescent   N
Piperaceae
Piper sanctum (Miq.) Schltdl. ex C. DC.   Shrubby   N
Punicaceae
Punica granatum L.   Arborescent   I
Rhamnaceae
Karwinskia humboldtiana (Schult.) Zucc.   Arborescent   N
Karwinskia latifolia Standl.   Arborescent   N
Rosaceae
Prunus persica (L.) Batsch   Arborescent   I
Prunus serotina Subsp. *Capuli* (Cav.) McVaugh   Arborescent   N
Rubiaceae
Coffea arabica L.   Shrubby   I
Genipa sp.   Arborescent   N
Randia capitata DC.   Shrubby   N
Randia laevigata Standl.   Shrubby   N
Rutaceae
Casimiroa sapota Oerst.   Arborescent   N
Casimiroa watsonii Engl.   Arborescent   N

Citrus aurantifolia Swingle	Arborescent	I
Citrus aurantium L.	Arborescent	I
Citrus limetta Risso	Arborescent	I
Citrus limon (L.) Osbeck	Arborescent	I
Citrus sinensis (L.) Osbeck	Arborescent	I
Sapotaceae		
Manilkara zapota (L) P. Royen	Arborescent	N
Mastichodendron capiri (A. DC.) Cronquist	Arborescent	N
Pouteria campechiana (Kunth) Baehni	Arborescent	N
Pouteria sapota (Jacq) H.E.Moore & Stearn	Arborescent	N
Sideroxylon cartilagineum (Cronquist) T.D. Penn	Arborescent	N
Sideroxylon persimile subsp. *subsessiliflorum* (Hemsl.) T.D. Penn.	Arborescent	N
Verbenaceae		
Lantana camara L.	Shrubby	N

Origin: N= Native, I=Introduced
The list was prepared by the authors

Most of these species are arboreal (77 percent) and the remaining (23 percent) are shrubs. As for their origins, 85 are native, and 11 are cultivars that were introduced into the BRS long ago. The distribution of these species in situ varies considerably due to temperature and rain patterns, as the eastern area of the BRS is drier and cooler than the western area, but the variation is mainly because of its usage record. In addition, the anthropogenic stress has been different in each area. In this regard it is fair to say that the greatest threat faced by these species has been the change in land use, mainly on account of livestock farming and the advance of agricultural frontiers involving excessive use of agrochemicals, in spite of state regulations from the Project of Ecological Land Management of the State of Jalisco,[5] which ban such land use in the study area and regard it as a protected area of environmental resources (Periódico Oficial del Estado de Jalisco 2001: 116).

Of the 96 arboreal and shrub species reported as phytogenetic resources for nourishment, 85 were acknowledged as such by the local residents of the study area. The remaining 11 are present in the site and might as well be eaten. Most edible species in the BRS (61 percent) are considered to have at least one additional use; some of them have even four other uses. Besides their edible use, they can have aesthetic or medicinal value, and can be used for forage or as living barriers. Furthermore, knowledge about the use of plants is heterogeneous in the different areas of the BRS, mainly considering that nowadays the few young people who still live there are not acquainted with them, in comparison with the knowledge that older people still have about plants.-

The 31 percent of the 96 species reported in the BRS are acknowledged by the United Nations' Food and Agriculture Organization (2012: 127, 285) as 'priority species for the reforestation in Mexico', and as shown in Table 5.2, they were selected for their ecological, economic and social value. Half of these species (15) are also shown in Table 5.2 to emphasise the significance of their environmental service and their social value. In addition, 9 species are noted for their importance to food security, and 8 of them can contribute to poverty reduction.

However, the BRS is also host to non-woody plants that belong to the so-called DTF habitat, which are also exploited as food sources. Although our census of these species is incomplete, the number as of today is 172 species, of which 74 are arborescent, 22 shrubs and 76 herbaceous. Nevertheless, the study zone is undergoing a process of reduction and degradation of its vegetable cover that entails a significant degree of genetic – and, in a sense, cultural – erosion of its wild vegetable resources.

Dietary Use of the Plants

As is the case throughout Mexico, the dietary base of the people who inhabit the BRS had its origins in pre-Hispanic times. Maize (*Zea mays*), beans (*Phaseolus vulgaris*), nopal (*Opuntia* spp.) and chilli peppers (*Capsicum annuum*) form part of the population's primordial sustenance, complemented, especially in rural and peri-urban areas, by the consumption of wild products and of plants grown in traditional gardens. We discovered one example of this in Ixcatán, a town in the BRS, whose residents still consume a basic alimentary package of pre-Hispanic origin but complement their alimentation by eating twenty-one wild edible species that we identified during the course of our study. Of these, nine are arboreal species characterised by seasonal production for a few months each year; and the commercialisation of 30 percent of this provides an important economic income for those families who are engaged in its collection for about four months. The other twelve species are herbaceous in nature, growing only during the rainy season and consumed in situ, though some have significant commercial potential (Sandoval Lozano 2015: 31).

A relevant case in BRS is the capomo (*Brosimum alicastrum*), a plant widely known as an alimentary species throughout the Mesoamerican territory. And yet, Pennington and Sarukán (2005: 138) did not report its existence in their estimated species distribution for the specific BRS area. During our fieldwork we completed more than twenty new recordings of this plant, which provided an idea of the breadth of its distribution beyond the limits already established by Pennington and Sarukán (2005:139); this is partly due to the lack of botanical collections of the species in the study area. Another significant case is that of *Malpighia mexicana*, a species related to the cultivation of acerola (*M. glabra, M. emarginata*) which is considered the most important natural source of ascorbic acid; in a preliminary analysis conducted by our account, we found vitamin C values for *M. mexicana* that were similar to commercial species. Although the plant is well represented in the BRS, it is no longer consumed.

Despite these precedents, however, it is clear that consumption of wild plants by townspeople in the BRS is declining steadily. As is the case almost everywhere else, this reduction in the consumption of traditional wild products, that are potentially sustainable and have various alimentary properties, is primarily due to cultural factors; that is, changes in eating patterns and the growing tendency to include industrially processed food products in daily diets, even though they are usually only of mediocre alimentary quality – and some may even be harmful to human health.

On the themes of health and nutrition, much remains to be resolved in Mexico. During the period, 2011–2012, chronic infantile malnutrition and acute malnutrition in children under five was at 13.6 percent and 1.6 percent respectively, well above the average for Latin America and the Caribbean which for 2015 was 11.3 percent and 1.3 percent respectively. During the same period, 2011–2012, the prevalence of overweight children under five was at 9 percent, well above the average for Latin America and the Caribbean which was 7.2 percent for 2014. For the year 2014, worrying levels of overweight scores (64 percent) and obesity (28 percent) were recorded in the adult (over 18 years of age) population. Consumption of ultra-processed foods (with high content levels of sugar, fat and salt) of lower nutritional value constituted one of the main factors that caused this increase in weight. In 2013, Mexico ranked high among the thirteen countries of Latin America and the Caribbean as regards the sale of ultra-processed foods and drinks, with 212.2 kilos per person. The high level of sales of these products relates to the lack of regulation of markets, rapid urbanisation and adoption of modern ways of life, and all the marketing. All of this caused an increase in the disponibility and disequilibrium of ultra-processed products, and a preference for their consumption (Food and Agriculture Organization, Pan American Health Organization 2017: 87, 88, 90-92, 96, 105). This situation was not much different in the populations of the barranca, where once avenues of communication had opened, and before the arrival of education and health services, the distribution and sale of fizzy drinks, beers, fried food and processed cookies took a hold.

Discussion and Conclusion

The ninety-six species of trees and shrubs, reported to contain one or more structures used for food in the BRS, are indicative of the potential of the region regarding food production. The fact that 88.5 percent are native to the area is also an indication that these might be within the ethnobotanical

Table 5.2 Edible trees and shrubs in the Barranca del Río Santiago considered primary to reforestation in Mexico because of their social and environmental service, and their use for food security and poverty alleviation.

Priority Species	Reasons for Prioritisation (Importance)	Environmental services or social value (code)*	Use for food security	Use for the reduction of poverty
Acacia farnesiana L. Willd.	Ecological and social	1; 2; 5; 7; 8;10; 11; 12	X	X
Annona cherimola Mill.	Ecological and social			
Annona reticulata L.	Ecological and social			
Annonna muricata L.	Ecological and social	7; 8; 9	X	X
Brosimum alicastrum Sw.	Ecological, economic and social	1; 3; 8; 10; 11; 12	X	X
Carica papaya L.	Ecological and social	5; 8		
Casimiroa sapota Oerst.	Ecological and social			
Ceiba aesculifolia (Kunth) Britten & Baker f.	Ecological and social			
Crescentia alata Kunth	Ecological and social	5; 10; 11		
Diospyros digyna Jacq.	Ecological, economic and social			
Enterolobium cyclocarpum (Jacq.) Griseb.	Ecological, economic and social	1; 2; 5; 7; 8; 10; 12	X	X
Genipa sp.	Ecological, economic and social	1; 2; 5; 10;11		
Guazuma ulmifolia Lam.	Ecological, economic and social	1; 2; 3; 5; 7; 10; 11;12	X	X
Inga vera Willd.	Ecological, economic and social	1; 5; 7; 10; 11		
Jacaratia mexicana A. DC.	Ecological and social			

Scientific name	Type of function	Functions*		
Juglans regia L.	Economic			
Leucaena esculenta (Moc. & Sessé ex DC.) Benth.	Ecological and social		X	
Leucaena leucocephala Lam.	Ecological, economic and social	1; 2; 5; 7; 8; 11; 12	X	X
Manilkara zapota (L) P. Royen	Ecological, economic and social	5; 8; 11		
Morus celtidifolia Kunth	Ecological and social			
Pithecellobium dulce (Roxb.) Benth.	Social	1; 2; 5; 7; 8; 10; 11		
Pouteria campechiana (Kunth) Baehni	Ecological and social			
Pouteria sapota (Jacq.) H.E.Moore & Stearn	Ecological and social			
Prosopis laevigata (Humb. & Bonpl. Ex Willd.) M.C.Johnst.	Ecological and social		X	
Prunus serotina Subsp. *Capuli* (Cav.) McVaugh	Ecological and social	1; 3; 5; 8; 10; 11; 12		
Psidium sartorianum (O.Berg) Nied.	Ecological and social			
Psidiuma guajaba L.	Ecological, economic and social	1; 2; 5; 7; 8; 11; 12		
Spondias mombin L.	Ecological and social			
Spondias purpurea L.	Ecological and social			
Tamarindus indica L.	Social	2; 5; 11; 12	X	X

*1=Soil and water preservation including catchment areas; 2=Soil protection; 3=Biodiversity preservation; 4=Cultural values; 5=Aesthetic values; 6=Religious values; 7=Fertility enhancement; 8=Soil recovery; 9=Dune fixation; 10=Live barriers; 11=Shadow; 12=Wind barrier.

Source: Authors' elaboration with data from Food and Agriculture Organization (2012).

knowledge that the BRS inhabitants possess about their food resources. It is evident that the potential of these phytogenetic resources for food is not identical for all these species, but for at least 31 percent of them there is an 'official' recognition in reforestation in Mexico because of their ecological and social importance, their environmental benefits and the social value that they represent, as well as because of their potential for food security and their contribution in the reduction of poverty.

Many of the species treated here correspond to traditional crops that now have high commercial value internationally (e.g. *Carica papaya* L., *Persea americana*). There are also introduced species from which specific cultivars have been generated for the area (e.g. *Mangifera indica*), as well as marginalised crops (such as *Annona* spp., *Pouteria* spp., *Spondias* spp.) and a large number of wild species capable of generating new cultivars (e.g. *Malpighia mexicana*), and many others used as seasonal food that can provide food in times of shortages.

Biodiversity for food and agriculture encompasses the components of biological diversity that are essential for feeding human populations and improving the quality of life, while the genetic resources for plants are the raw material on which the world depends in order to improve the productivity and quality of crops (Food and Agriculture Organization 2018a, 2018b). Hence, it is a priority to give greater attention to the conservation of wild plant species for direct, indirect or potential use. Yet, these are threatened by the change of land use, by the use of herbicides and the increase of harmful substances in the atmosphere, putting genetic diversity at risk by promoting the disappearance of useful or usable wild species (Esquinas Alcázar 1983: 23). Furthermore, the Food and Agriculture Organization recognises that, over the past one hundred years, agricultural crops worldwide have lost 75 percent of their genetic diversity (Food and Agriculture Organization 2005: 3).

On a global scale, studies indicate that the loss of biodiversity does not slow down. On the contrary, it will continue to increase if additional public policies related to rural development and better food availability are not implemented (Netherlands Environmental Assessment Agency 2010). An example is that modern agriculture and forestry have caused, until 2010, almost 60 percent of the total reduction of terrestrial biodiversity. Furthermore, according to the Mean Species Abundance, by 2050 the expected loss of biodiversity will be in the region of 55 percent (Kok et al. 2018: 136).

Tragically, although Mexico is still classified, in principle, as a 'mega-diverse' country, its biodiversity is disappearing fast. DTF once covered approximately 14 percent of the nation's territory (some 280,000 km^2), whereas today, over three-quarters of that surface area (around 210,000 km^2) has been converted to alternative uses, including secondary vegetation, agricultural land, dams, urban, industrial and tourist developments, and diverse means of communication. In fact, in some regions of the country, the natural vegetation (i.e. not affected by humans – the so-called pristine type) has been lost completely (Rzedowsky and Calderón 2013: 4–5).

In relation to this, it becomes opportune to refer here to Villaseñor, who has pointed out:

Mexico has a long and growing tradition of studying its vascular flora, reflected in the significant increase in recent decades of specimens housed in national scientific collections and abroad, backed by an immense bibliography. However, the knowledge of national floristic richness is still unsatisfactory mainly due to the difficulty of synthesising scattered information in such publications along with the lack of well-curated databases of specimens documenting this richness. (Villaseñor 2016: 559)

Furthermore, Luna and colleagues (2011) write:

Because of these circumstances, Mexico now often appears on the 'Red Lists' compiled by the International Union for Nature Conservation; it ranks first in Latin America as regards endangered species, of which there are 897 species (Luna et al. 2011: 38). This means that the source of ... foods, medicines and potable water, as well as the means of subsistence of millions of people, could be at risk due to the rapid reduction of the world's animal and vegetable species (Unión Internacional para la Conservación de la Naturaleza 2019).

Unfortunately, we do not know with any precision just what the state of biodiversity is in Mexico today. The paucity of information on the vegetable cover of its territories impedes making any detailed comparative comments (Flores-Vilella and Gerez 1994: 1; Rzedowsky 2006: 16). In fact, only a few biological groups have been diagnosed in any depth in this regard. So, it is worth mentioning a paper by Vega (2001) in which he reviews the Mexican economy for a period of sixty years (1940–2000). He demonstrates the unsustainability of economic and social development in Mexico, and points out that the common marker, whether it is a period of material well-being or of economic slowdown, has been the concentration of national income and the sum total of 'environmental passive'.[6] This means that the periods of economic expansion, as well as the recessive ones, correspond to cases of social exclusion and 'severe ecological degradation processes, depletion of natural resources and environmental contamination' (Vega 2001: 31). In Mexico the costs of environmental protection represent 1 percent of the *producto interno bruto* (P.I.B.) [gross domestic product] P.I.B., whereas the official estimate of the total costs of environmental degradation corresponds to 5.7 percent of P.I.B. In other words, the country is losing six times more due to environmental degradation than it is investing in preventing, controlling, reducing or compensating for environmental degradation or managing the protection of the environment (Comisión Nacional para el Conocimiento y Uso de la Biodiversidad 2016: 156). Moreover, the Food and Agriculture Organization has found that Mexico lacks public policies to foster the study and evaluation of the genetic variation of its woody species, and has failed to develop mechanisms to monitor and assess such problems as the vulnerability of plants and their genetic erosion. 'Extreme poverty, environmental degradation and loss of natural resources are highlighted among priorities that require immediate attention in our country. To the extent that these priorities are addressed and resolved, the preservation and sustainable use of natural resources, based on forestry, favourable changes in equity issues, and insecurity will be promoted' (Food and Agriculture Organization 2012: 3, 109).

Although in the BRS the native vegetation has been transformed by anthropocentric activity, and agricultural production has become permeated by the technologies of the Green Revolution with little implementation of sustainable or environmentally friendly techniques, nevertheless the area has much to contribute as regards plant genetic resources for food.

In summary, constructing alimentary alternatives for sustainability in the twenty-first century requires a great deal of high-quality research as well as proposals for innovation and development. The issues of conservation and restoration of biodiversity, as well as the sustainable use of resources, must be an integral part of sustainable development strategies, which in terms of primary production concern agriculture, forestry, fisheries and energy (Kok et al. 2018: 137). In the Mexican case, the state, private entities and institutions devoted to scientific research and higher education, must join forces to supply the material and intellectual resources required: (1) to conduct rigorous, systematic studies that assess the current condition of the nation's vegetable resources; (2) to elaborate prognoses of their future evolution; (3) to propose changes in land use; and (4) to design public policies that will guide the exploitation and conservation of wild phytogenetic resources.

Martín Tena Meza
Universidad de Guadalajara, Mexico

Rafael M. Navarro-Cerrillo
Universidad de Córdoba, Spain

Ricardo Ávila Palafox
Universidad de Guadalajara, Mexico

Raymundo Villavicencio García
Universidad de Guadalajara, Mexico

Notes

1. www.fao.org/plant-treaty/overview/en/ (last accessed 14 February 2019).
2. www.fao.org/nr/cgrfa/cgrfa-global/cgrfa-globplan/en/ (last accessed 14 February 2019).
3. These findings are part of the PhD research of the lead author as part of the doctoral programme of Bioscience and Agro-food Sciences of Córdoba University, Spain.
4. The terminology used for edible species or edible plants accounts for any type (weed, bush or tree) with one or more structures used for alimentary purposes.
5. In Spanish: *Proyecto de Ordenamiento Ecológico del Territorio del Estado de Jalisco*.
6. An 'environmental passive' refers to a company's debt regarding the environmental damage it has caused.

References

Casas, A. (2010) Prólogo. In Lascurain, M., Avendaño, S., del Amo, S. and Niembro, A. *Guía de frutos silvestres comestibles en Veracruz*, Fondo Sectorial para la Investigación, el Desarrollo y la Innovación Tecnológica Forestal, Conafor-Conacyt, Mexico, pp. 7–9.

Challenger, A. (1998) *Utilización y conservación de los ecosistemas terrestres de México: Pasado, presente y futuro*, Universidad Nacional Autónoma de México, Instituto de Biología, Mexico, p. 847.

Comisión Nacional para el Conocimiento y Uso de la Biodiversidad (2016) *Estrategia nacional sobre biodiversidad de México y Plan de acción 2016–2030*, Mexico, p. 384.

Esquinas Alcázar, J. (1982) *Los recursos fitogenéticos una inversion segura para el futuro*, Consejo Internacional de Recursos Fitogenéticos, Food and Agriculture Organization, p. 44.

Flores-Vilella, O. and Gerez, P. (1994) *Biodiversidad y conservación en México: vertebrados, vegetación y uso de suelo*, Second Edition, Comisión Nacional para el Conocimiento y Uso de la Biodiversidad, Universidad Nacional Autónoma de México, Mexico.

Food and Agriculture Organization (2005) *Building on Gender, Agrobiodiversity, and Local Knowledge – A Training Manual*, Rome.

Food and Agriculture Organization (2012) *Forest Genetic Resources Situation in Mexico*, Rome, p. 288.

Food and Agriculture Organization (2018a) Home page topic Biodiversity. Published at: http://www.fao.org/biodiversity/pagina-principal-de-biodiversidad/es/. Accessed on 12 April 2018.

Food and Agriculture Organization (2018b) Home page topic Genetic Resources. Published at: http://www.fao.org/genetic-resources/es/. Accessed on 12 April 2018.

Food and Agriculture Organization, Pan American Health Organization (2017) *Panorama de la seguridad alimentaria y nutricional 2016 – América Latina y el Caribe*, Santiago, Chile, p. 163.

Kirchhoff, P. (1967) Mesoamérica: Sus límites geográficos, composición étnica y caracteres culturales, *Suplemento de la Revista Tlatoani*. Number 3, ENAH-INAH, Mexico, p. 13.

Kok, M., Alkemade, R., Bakkenes, M., van Eerdt, M., Janse, J., Mandryk, M., Kram, T., Lazarova, T., Meijer, J., van Oorschot, M., Westhoeka, H., van der Zagtc, R., van der Berg, M., van der Esch, S. Prins, A. and van Vuuren, P. (2018) Pathways for Agriculture and Forestry to Contribute to Terrestrial Biodiversity Conservation: A Global Scenario-Study, *Biological Conservation*, 221: 137–150.

López, R., Cházaro, M., González, R. and Covarrubias, H. (2011) Árboles de las barrancas de los Ríos Santiago y Verde, Comisión Estatal del Agua de Jalisco, Mexico.

Luna, R., Castañón, A. and Raz-Guzmán, A. (2011) La biodiversidad en México su conservación y las colecciones biológicas, *Ciencias*, 101: 36–43.

McVaugh, R. (1972) Botanical Exploration in Nueva Galicia, Mexico from 1790 to the Present Time, *Contribution from the University of Michigan Herbarium* 9(3–7): 205–358.

Molina, M. and Córdova, T. (eds) (2006) *Recursos Fitogenéticos de México para la Alimentación y la Agricultura: Informe Nacional 2006*, Secretaría de Agricultura, Ganadería, Desarrollo Rural, Pesca y Alimentación y Sociedad Mexicana de Fitogenética, A.C. Chapingo, Mexico, p. 174.

Netherlands Environmental Assessment Agency (2010) Rethinking Global Biodiversity Strategies, The Hague/Bilthoven. Published at: https://www.pbl.nl/en/publications/2010/Rethinking_Global_Biodiversity_Strategies. Accessed on 20 January 2019.

Pennington, T. and Sarukhán J. (2005) Árboles tropicales de México: *Manual para la identificación de las principales especies*, Fondo de Cultura Económica, Mexico, p. 523.

Periódico Oficial del Estado de Jalisco (2001) 28 July. *Ordenamiento Ecológico Territorial del Estado de Jalisco*, Secretaría de Medio Ambiente para el Desarrollo Sustentable, Gobierno del Estado de Jalisco, Mexico, p. 192.

Rzedowsky, J. (2006) *Vegetación de México* (First digital edition), Comisión Nacional para el Conocimiento y Uso de la Biodiversidad, Mexico, p. 504.

Rzedowsky, J. and Calderón, G. (2013) Datos para la apreciación de la flora fanerogámica del bosque tropical caducifolio de México, *Acta Botánica Mexicana*, 102: 1–23.

Sahagún-Godínez, E., Macías Rodríguez, M.-A., Carrillo Reyes, P. and Vázquez García, J.A. (2014) *Guía de campo de los arboles tropicales de la Barranca del Río Santiago en Jalisco, México*, Prometeo Editores, Mexico.

Sandoval Lozano, C. (2015) *La Cultura alimentaría de San Francisco de Ixcatan, Jalisco*, Tesis Licenciatura en Biología, Universidad de Guadalajara, Mexico, p. 107.

Secretaría de Medio Ambiente y Desarrollo Territorial (2016) *Estudio Técnico justificativo y Programa de manejo del Proyecto de Declaratoria del Área Natural Protegida Formación Natural de Interés Estatal Barranca de los Ríos Santiago y Verde*, Jalisco, Mexico, p. 389.

Soto, J. (2010) Plantas útiles de la cuenca del Balsas. In Ceballos, G., Martínez, L., García, A., Espinoza, E., Bezaury, J. and Dirzo, R. (eds) *Diversidad, amenazas y áreas prioritarias para la conservación de las selvas secas del Pacífico de México*.

Fondo de Cultura Económica / Comisión Nacional para el conocimiento y uso de la Biodiversidad, Mexico, pp. 285–320.

Trejo, I. (2005) Análisis de la diversidad de la Selva baja caducifolia en México. In Halffter, G., Soberón, J., Koleff, P. and Melic, A. *Sobre Diversidad Biológica: El significado de las Diversidades Alfa, Beta y Gamma*, Monografías Tercer Milenio, pp. 111–122.

Unión Internacional para la Conservación de la Naturaleza (2019) *Lista Roja de UICN*. Published at https://www.iucn.org/es/regiones/am%C3%A9rica-del-sur/nuestro-trabajo/pol%C3%ADticas-de-biodiversidad/lista-roja-de-uicn. Accessed on 14 February 2019.

Universidad de Guadalajara (2009) Protejamos la Barranca del Río Santiago. Extención Universitaria Noticias. Published at: http://udg.mx/es/noticia/protejamos-la-barranca-del-rio-santiago. Accessed on 20 January 2019.

Vargas-Ponce, O., Zizumbo, D. and Colunga, P. (2007) In Situ Diversity and Maintenance of Traditional Agave Landraces Used in Spirits Production in West-Central Mexico, *Economic Botany*, 61(4): 362–375. DOI: 10.1663/0013-0001(2007)61[362:ISDAMO]2.0.CO;2.

Vega, E. (2001) La sustentabilidad en México: ¿estamos mal pero vamos bien? *Gaceta Ecológica*, 61: 30–45.

Villaseñor, J.L. (2016) Checklist of the Native Vascular Plants of Mexico. *Revista Mexicana de Biodiversidad*, 87(3): 559–902. Published at: http://www.revista.ib.unam.mx/index.php/bio/issue/view/50/showToc. Accessed on 4 August 2016.

Zizumbo-Villarreal, D., Flores-Silva, A. and Colunga-GarcíaMarín, P. (2014) The Food System during the Formative Period in West Mesoamerica, *Economic Botany*, 68: 67–84. DOI 10.1007/s12231-014-9262-y.

CHAPTER 6
FARM URBAN AND URBAN AQUAPONICS: CHANGING PERCEPTIONS IN CLASSROOMS AND COMMUNITIES

Iain Young

Rising food and energy prices, increasing unemployment and unhealthy, unsustainable lifestyles are major concerns for today's society. Aquaponics offers the possibility of increasing food sustainability by lowimpact, high-density agriculture at the domestic and community level, reducing transport-related carbon emissions and promoting resilience in the face of supply chain interruptions (Kotzen 2013; Goddek et al. 2015; Cunningham and Kotzen 2015).

Aquaponics combines soil-less growing (hydroponics) with fish husbandry (aquaculture) in an enclosed recirculation system. The combination of growing fish and plants together is not new. Early examples include artificial islands or 'floating gardens' in shallow lakes in central America, where plants were fertilised using the mud and waste grudged from the lake bed (e.g. Aztec's Chinampas 1350–1150 BC) (Turcios and Papenbrock 2014). In this example, however, the presence of wild fish is more or less coincidental rather than designed into the system. Aquaculture of fish in rice fields in South East Asia dates from around 1,500 years ago, and more recently fish have been grown out with rice and other grain production around the world (Coche 1967; Ahmad 2001). A pond, garden and livestock system called the 'vuon, ao, chuong' or 'VAC' was promoted widely in Vietnam in the early 1980s (Lee Thanh Luu 2001). It is generally accepted that the latest incarnation of aquaponics started in the late 1970s and early 1980s.

Linking aquaculture and hydroponic production in a circular, synergistic way provides an opportunity to address any shortcomings with each. This closed nutrient cycling fulfils the definition of sustainability offered by Lehman et al. (1993), who define sustainable agriculture as 'a process that does not deplete any non-renewable resources'. Similarly, the cycling nature of the aquaponics system is close to that described by Francis et al. (2003) as a production system resembling a natural ecosystem, with 'closed nutrient cycles'. Further benefits are gained from reuse of the water via recirculation.

In a simple form of aquaponics, the nutrient-rich waste water from the fish tanks fertilises hydroponic beds. The plants and their associated bacteria remove these nutrients from the water. Uneaten fish feed, fish waste, algae and other organic material in the fish tank water all contribute to this pool of nutrients, acting as fertiliser for the hydroponically cultivated plants. In turn, the hydroponic beds act as a biofilter, stripping toxic ammonia out of the water by providing a suitable environment and large surface area for the ammonia oxidising 'nitrosomonas' bacteria and the nitrite oxidising 'nitrobacter' bacteria. The cleansed water is then recirculated back into the fish tanks.

Researchers, including Mark McMurtry and Doug Sanders at North Carolina State University (Love et al. 2014), and James Rakocy at the University of the Virgin Islands, developed modern aquaponics systems (Rakocy 1989). Mark McMurtry described the aquaculture system at North Carolina State University as an 'Integrated Aqua-Vegeculture System' (McMurtry 1988) consisting of tilapia tanks connected to sand hydroponic vegetable beds. Rakocy at the University of the Virgin Islands developed deep-water aquaponics, where plants are cultured on rafts floating in deep-water troughs circulated with water from fish tanks. McMurtry and, separately, Rakocy note that these aquaponics systems are suitable for use in arid and semi-arid regions where water is scarce and demand for protein and fresh vegetables is high. More recently, there has been a growing interest in aquaponics as an innovative solution to food security (Love et al. 2014). Diver (2006) identified John and Nancy Todd's 'The New Alchemy Institute' on Cape Cod as the earliest developer of integrated fish and vegetable systems. From 1971 to 1991, the institute was 'a rigorous research farm investigating closing the loop between food production, the built environment and waste'[1] . In an important step in the development of modern aquaponics, the New Alchemy Institute developed the 'solar pond' (or solar aquaculture), using the thermal mass of above-ground fish tanks to maintain a more constant temperature in greenhouses (Zweig 1986). A floating hydroponic system was added to this, separated from the fish with a mesh barrier, so that tilapia or catfish could be grown alongside, but separated from, lettuce.

Diver (2006) credits the owners of S&S Aquafarm, Tom and Paula Speraneo, with the evolution of aquaponics into commercial-scale production in the early 1990s. A substantial system of pea-gravel beds and tilapia tanks was used to grow 'fresh basil, tomatoes, cucumbers, mixed salad greens and an assortment of vegetable, herb, and ornamental bedding plants in the aquaponic greenhouse'. The Speraneos were keen proponents of aquaponics; they produced training materials and opened up the site of their commercial-scale system to visitors. More than ten thousand visitors of all ages visited over its life. Like the S&S Aquafarm, the Freshwater Institute in Shepherdstown, West Virginia, built demonstration aquaponics systems and produced training materials focused on the design and operation of aquaponics systems, including reciprocal hydroponics where the growing media are periodically flooded and drained to supply oxygen and water to the plants.

Rakocy and colleagues from the University of the Virgin Islands Agricultural Experiment Station utilised deep-water raft hydroponics, similar to New Alchemy's 'solar pond'. The system was sophisticated, with filters, oxygenation systems, temperature and pH control systems, and their research focused primarily on the more technical aspects of aquaponics, from nutrient balance to system design (Rakocy 1999a, 1999b, 2007). Rakocy also worked with the commercial operation of Nelson and Pade Inc.[2] to develop training and educational materials. Nelson and Pade market aquaponics systems and supplies, planning services, educational materials and training courses, primarily in the United States.

The growing popularity of aquaponics may be fuelled more by social and environmental benefits and the promotion of a sustainable lifestyle, than by cold economics. Indeed, there have been very few appraisals of the financial costs and benefits of commercial aquaponics. Possibly the first quality review, including analysis of over 250 commercial aquaponics operations, was carried out by Love et al. (2015). They comment that the growing demand for locally produced food sold directly to consumers, has helped to promote the commercialisation of aquaponics. However, the most successful operations are those that diversify their revenue streams beyond the sale of aquaponics production crops by selling non-food products (e.g. aquaponics systems and supplies), services and training courses. The last decade has seen a growing number of aquaponics training courses and education programmes, often provided by private companies but also by community interest groups and educational institutions. The commercial courses tend to be well marketed, utilise quite sophisticated systems such as those run by 'Humble by Nature', utilising a sophisticated 'Passive Solar Greenhouse' and by 'Bioaqua farm' in their commercial Aquaponics Farm[3]. These courses promote a resort ethos, providing meals using the produce from the farm and including a social calendar.

Education Using Aquaponics

Engagement with nature has benefits beyond the development of Science, Technology, Engineering and Mathematics (STEM) knowledge and skills, because understanding the financial, environmental and social benefits of local production of food and improved access to fresh fish and vegetables provides insight into environmental sciences, economics, marketing and nutrition. Furthermore, there are other less tangible benefits. Barton and Pretty (2010) investigated the impact of 'nature' on self-esteem and mood, and they were able to prove the positive influence of experiences of nature on such parameters. The extent of the influence was found to depend on participant age and the duration, intensity and location of the experience. A five-minute active experience of nature (compared to longer periods including 10–60 minutes, half- and full-day experiences) had the strongest effect on self-esteem and mood, and the most intense influence was seen in people under thirty years of age.

While there is little doubt of the value of education in rural environments, these opportunities can be costly in terms of both time and money. However, such experiences can be especially valuable in, or close to, cities, and can be provided at low cost. The 'Green Classroom' in the Botanical Garden of the University of Ulm is visited by about 2,500 school-age students a year. It uses experiential learning to expand students' biological knowledge and develop positive attitudes towards animals and plants. Students who have visited the Green Classroom show a better understanding of biology and ecosystems than their peers (Drissner et al. 2011). Utilising the whole university campus to provide a botanical tour, Ratnayaka (2017) showed that the outdoor, hands-on, experiential learning (botanical tour) increased student satisfaction and motivation, leading to a deeper understanding of the subject compared to instruction-based learning in a classroom. Student perception and their results in graded exercises improved by around 10 percent using active-learning compared to traditional classroom plant-related laboratories (Ratnayaka, 2017).

Aquaponics provides an attractive educational tool. The systems contain plants, fish and bacteria; they are self-contained ecosystems and they allow teachers and students to explore a wide range of science, technology, engineering and mathematics topics (Schneller et al. 2015; Genello et al. 2015). Even in a classroom far from the countryside, students can engage with nature in a very hands-on way, caring for the plants and animals, performing water tests, feeding the fish, planting, tending and harvesting the plants.

Education, Aquaponics and Urban Farming Projects in and around Liverpool

Farm Urban

Farm Urban was founded by two bio-science graduates, Paul Myers and Jens Thomas, from the University of Liverpool. They set themselves a goal of using their biological background and their interest in growing and eating healthy food to find solutions to food security challenges by 'linking leading scientific research with local food production' and 'developing and testing the most efficient ways to grow food in urban environments'. Early in their story, Farm Urban recognised that aquaponics and urban food production provide a very effective focal point for education and engagement programmes. So, by focusing on sustainable urban living for local communities and by engaging with schools, communities, hospitals and universities, they started to develop and promote education programmes to demonstrate how anyone can produce healthy food in a sustainable and cost-effective way. They also wanted to draw attention to the wider issues around food security. Collaborating with the University of Liverpool and the Liverpool Life Sciences University Technical College, they developed workshops, STEM clubs and enrichment activities focused on a Produce Pod – a 'hacked' aquaponics system so that the secondary school students' STEM clubs, using the produce pods, can participate in collaborative research providing and using crowd-sourced data and participating in online forums[4].

University of Liverpool Guild Farm

The University of Liverpool Guild of Students set up the 'Green Guild' to encourage students not just to think about, but to get involved in 'green' and 'sustainable' activities. The Green Guild organises student-led volunteering initiatives and social enterprise projects as well as finding funding for its own initiatives such as 'Student switch off' – a competition between university halls of residence to save water and energy and increase recycling. Green Schools, in collaboration with the University Widening Participation team, places volunteers in local schools to promote sustainability and enterprise skills. The Green Guild also purchased a rapid composter and processes all of the food preparation waste, including, for example, the grinds from the in-house coffee shop, to create compost for their own roof garden and for other projects. The Green Guild also set up a Social Enterprise Challenge which, together with a generous grant from the 'Friends of the University of Liverpool', funded Farm Urban to design and install an 'urban farm' on the roof of the Guild of Students' Building to demonstrate aquaponics growing techniques in a 'typical' disused urban space. The farm is maintained by Farm Urban and a group of volunteers and is used for education and outreach events to engage with school students and other members of the local community.

Alder Hey Children's Hospital

The Alder Hey Children's Hospital has just been through a quarter of a billion pound redesign to create one of the most innovative and environmentally sustainable hospitals in the world. The young patients were involved in every step of the design process, creating a fun, child- and family-centred environment in an eco-friendly building that generates a portion of its own energy, has 'green roofs' planted to encourage biodiversity and part-indoor, part-outdoor 'playdecks' so that even sick children can benefit from the feel of being outside while being kept safe from the elements. Funded by a legacy from a local Liverpool resident, Farm Urban designed and installed three bespoke aquaponic systems on the playdecks. The water from Koi and goldfish in the fish tanks feeds plants and vegetables in racks above, which in turn clean and filter the water. These systems provide a focal point for patients, parents, staff and visitors who enjoy feeding the fish and learning about the mini-ecosystem. Furthermore, the herbs and salad are used in the children's meals by hospital chefs so that they enrich the diet of the children and encourage them to think about the source of their food and the value of fresh, healthy vegetables in their diet. This project has featured heavily in the local and national media.[5]

University Technical College, Liverpool (UTC)

The Liverpool UTC has academy status; it is a life sciences technical college in the so-called 'creative quarter' of Liverpool, 'the Baltic Triangle'. Its students are aged 14–19 and are being taught in an innovative hands-on laboratory environment that encourages the early development of a wide range of professional skills and student-led critical learning. Farm Urban runs an aquaponics enrichment programme with UTC students each year, which has produced a DNA double-helix-inspired aquaponic system that is on prominent display in the college foyer[6]. Farm Urban is also installing a productive farm and laboratory in the college's basement. Aquaponics has been used at the school to provide a hands-on laboratory for life sciences, engineering and design students and to promote a dialogue around issues of global food security and its potential solutions.

Ness Gardens

Situated on the Wirral Peninsula, the award-winning Ness Gardens is Liverpool University's botanical garden and research facility. The gardens were created in 1898 by Arthur Kilpin Bulley from Liverpool[7]. Bulley funded plant-hunting expeditions to the Far East and established the Bees Seeds company, famous for its packets of cheap seeds, which were sold for decades in shops like Woolworths[8]. The gardens were donated to the University of Liverpool in 1948 by Bulley's daughter to use for research and education. It is a key visitor attraction on the Wirral. The Ness visitor centre is powered by an array of solar panels, it is topped with a green roof, and much of the food sold there is grown on site. Farm Urban transformed an overgrown conservatory attached to the visitor centre into an engaging installation showcasing different aquaponic growing techniques to inspire visitors.

Duke of York Young Enterprise Award

Farm Urban was awarded a prestigious Duke of York Young Enterprise Award in 2016. After reaching the final, the managing director, Paul Meyers, had to deliver a one-minute 'elevator pitch' to explain what Farm Urban is, who is involved and what they do. The award was a validation for their business approach, which is more about slow, organic growth while also developing skills and a network of collaborators, rather than seeking large amounts of capital investment.

Conclusion

Urban farming is unlikely to replace conventional agriculture; for example, it makes little sense to grow easily transportable foods like potatoes in cities. However, it should be complementary to it, growing, for example, high-demand, easily perishable, short shelf-life foods, which are consequently the foods that are highest on the charts of food that is wasted, such as salads. The first step is to educate the public in the benefits of healthy eating and local food production.

Iain Young
Institute of Clinical Sciences, University of Liverpool, UK

Notes

1. http://www.toddecological.com (accessed on 1 August 2017).
2. https://aquaponics.com/ (accessed on 2 August 2017).
3. http://www.humblebynature.com/courses-humble-by-nature/aquaponics (last accessed on 2 August 2017); http://bioaquafarm.co.uk/courses/ (accessed on 2 August 2017).
4. https://www.farmurban.co.uk/ (accessed on 29 January 2019).
5. https://www.thetimes.co.uk/article/couples-diamond-gift-puts-aquaponics-on-hospital-menu-20xrnbkhq (last accessed on 25 April 2018); http://www.bbc.co.uk/news/av/health-34411212/inside-liverpool-s-new-alder-hey-children-s-hospital (accessed on 25 April 2018).
6. http://www.farmurban.co.uk/projects (accessed on 25 April 2018).
7. http://www.nessgardens.org.uk/the-gardens/history (accessed on 3 August 2017).
8. http://www.independent.co.uk/arts-entertainment/a-place-where-ness-is-more-1156378.html (accessed on 3 August 2017).

References

Ahmad, R. Sh. Hj. (2001) Fodder-Fish Integration Practice in Malaysia. In *Integrated Agriculture-Aquaculture: A Primer,* Food and Agriculture Organization Fisheries Technical Paper 407, FAO, Rome, p. 33.

Barton, J. and Pretty, J. (2010). What is the Best Dose of Nature and Green Exercise for Improving Mental Health? A Multi-Study, *Environmental Science and Technology,* 44(10): 3947–3955.

Coche, A.G. (1967) Fish Culture in Rice Fields a Worldwide Synthesis, *Hydrobiologia,* 30: 1–44.

Cunningham, H. and Kotzen, B. (2015) Meet the Sustainable Vegetables That Thrive on a Diet of Fish Poo. Published at: http://theconversation.com/the-sustainable-vegetables-that-thrive-on-a-diet-of-fish-poo-50160. Accessed on 25 April 2018.

Diver, S. (2006) [Updated by Lee Rinehart (2010)] Aquaponics – Integration of Hydroponics with Aquaculture. NCAT Agriculture Specialists IP163. Published at: https://attra.ncat.org/attra-pub/viewhtml.php?id=56. Accessed on 25 April 2018.

Drissner, J., Haase, H.-M., Nikolajek, M. and Hille, K. (2011) Environmental Education in a 'Green Classroom', *Resonance,* 16(2): 180–187.

Francis, C., Lieblein, G., Gliessman, S., Breland, T.A., Creamer, N., Harwood, R., Salomonsson, L., Helenius, J., Rickerl, D. Salvador, R., Wiedenhoeft, M., Simmons, S., Allen, P., Altieri, M. Flora, C. and Poincelot, R. (2003) Agroecology: The Ecology of Food Systems, *Journal of Sustainable Agriculture,* 22: 99–118.

Genello, L., Fry, J.P., Frederick, J.A., Li, X. and Love, D.C. (2015) Fish in the Classroom: A Survey of the Use of Aquaponics in Education, *European Journal of Health & Biology Education,* 4(2): 9–20.

Goddek, S., Delaide, B., Mankasingh, U., Ragnarsdottir, K.V., Jijakli, H. and Thorarinsdottir, R. (2015) Challenges of Sustainable and Commercial Aquaponics, *Sustainability: Science Practice and Policy,* 7: 4199–4224.

Kotzen, B. (2013) Memorandum of Understanding for the Implementation of a European Concerted Research Action Designated as COST Action FA1305: The EU Aquaponics Hub: Realising Sustainable Integrated Fish and Vegetable Production for the EU. European Cooperation in Science and Technology (COST).

Lee Thanh Luu (2001) The VAC System in Northern Vietnam. In *Integrated Agriculture-Aquaculture: A Primer,* Food and Agriculture Organization Fisheries Technical Paper 407, FAO, Rome, p. 26

Lehman, H., Clark, E.A. and Weise, S.F. (1993) Clarifying the Definition of Sustainable Agriculture, *Journal of Agricultural and Environmental Ethics,* 6: 127–143.

Love, D.C., Fry, J.P., Genello, L., Hill, E.S., Frederick, J.A., Li, X. and Semmens, K. (2014) An International Survey of Aquaponics Practitioners, *PLoS One,* 9: e102662.

Love, D.C., Fry, J.P., Li, X., Hill, E.S., Genello, L., Semmens, K. and Thompson, R.E. (2015) Commercial Aquaponics Production and Profitability: Findings from an International Survey, *Aquaculture,* 435: 67–74.

McMurtry, M.R. (1988) *Aqua-Vegeculture Systems: International Ag-Sieve,* Vol. 1, No. 3. Rodale Press International, Emmaus, PA.

Rakocy, J.E. (1989) Hydroponic Lettuce Production in a Recirculating Fish Culture System, University of the Virgin Islands, Agricultural Experiment Station, *Island Perspectives,* 3: 4–10.

Rakocy, J.E. (1999a) Aquaculture Engineering: The Status of Aquaponics, Part 1, *Aquaculture Magazine*, 25(4).

Rakocy, J.E. (1999b) Aquaculture Engineering: The Status of Aquaponics, Part 2, *Aquaculture Magazine*, 25(5).

Rakocy, J.E. (2007) Aquaponics: Preliminary Evaluation of Organic Waste from Two Aquaculture Systems as a Source of Inorganic Nutrients for Hydroponics, *Acta Horticulturae*, 742: 201.

Ratnayaka, H.H. (2017) An On-campus Botanical Tour to Promote Student Satisfaction and Learning in a University-Level Biodiversity or General Biology Course, *Education Sciences*, 7(1): 1–12.

Schneller, A.J., Schofield, C.A., Frank, J., Hollister, E. and Mamuszka, L. (2015) A Case Study of Indoor Garden-based Learning with Hydroponics and Aquaponics: Evaluating Pro-environmental Knowledge, Perception, and Behavior Change, *Applied Environmental Education and Communication*, 14(4): 256–265.

Turcios, A.E. and Papenbrock, J. (2014) Sustainable Treatment of Aquaculture Effluents: What Can We Learn from the Past for the Future? *Sustainability*, 6: 836–856.

Zweig, R. (1986) *Aquaculture Magazine*: 34–40. Published at: https://newalchemists. files.wordpress.com/2015/01/an-integrated-zweig-text-4-photos1.pdf. Accessed on 24 April 2018.

CHAPTER 7
'DIG FOR SUSTAINABILITY' IN THE TWENTY-FIRST CENTURY: ALLOTMENTS, GARDENS AND TELEVISION

Helen Macbeth

Introduction

Has growing food for the family become an anachronism in Britain? This chapter argues that we can and should learn from the past, and, as regards support for food sustainability, it specifically refers to the food policies of the government of the United Kingdom in the Second World War and the nutritional state of the population as a result of these policies. The emphasis will be on the increased activity at that time by so many to grow vegetables on any land that could be found for planting, and even to raise a few chickens or a pig in urban and suburban settings, sometimes in cooperation between neighbours. To be discussed here is whether any of the factors that helped to maintain the nutrition of the British population in wartime could be relevant to food sustainability and nutritional standards in the economic situation of the early twenty-first century.

Food in Britain at War, 1939–1945

The British Isles have a large population; the island of Ireland was, at the time of the Second World War, and is still today, divided between Northern Ireland, a part of the UK, and the Republic of Ireland. The latter remained neutral during the war. Before 1939, the UK depended on imports for more than half of its supplies of food (Ministry of Food 1946). Imported food came from Europe and from across the world, all of necessity by sea, which is of course the basis of Britain's strong maritime history, with established trade patterns developed globally, and in particular within the British Empire. The success of the German U-boat fleet caused great losses to merchant shipping and thereby a huge reduction in imported foodstuffs. Furthermore, as European countries became allied with or were invaded by Germany, food imports from Europe ceased. Yet, nutrition of the population was maintained. The plentiful work of the Oxford Nutrition Surveys (Sinclair 1944) resulted in reports that mostly remain unpublished. Recent research on the contemporary 1940s medical surveys (Darke 1979; Huxley et al. 2000) support the claim that the nutritional state of the British population, reviewed across all socio-economic groups, was better during the war than before or after. Not surprisingly, that includes avoidance of the malnutrition of an excessive or over-indulgent diet.

Learning from the past, however, was also relevant to the preparations for war before 1939, as both sides in the First World War (1914–1918) had achieved partial sea blockades against the importation of essentials for the other side. In Britain in that war, there had been wartime inflation in the cost of food (Mitchell and Dean 1962), resulting in serious undernutrition in some sectors of the population, but also an increase in allotments, the original 'Dig for Victory' campaign and the launch of the Women's Land Army. So, in the late 1930s, the UK government, envisaging the likelihood of another war, undertook studies and made plans for survival to counter similar blockades; these plans were put into place immediately that war was declared in September 1939. A new 'Ministry of Food' was set up, and some remarkable individuals[1] were involved in the planning of food policies; their determination to create a holistic approach to the problem, while bearing in mind all the relevant perspectives, was highly significant for the success of their policies. The Ministry of Food worked in conjunction with the Ministry of Agriculture, and all aspects of food production, supply and distribution became government controlled (Lightowler and Macbeth 2014).

In brief overview, the objectives of the Ministry of Food in the Second World War were to maintain adequate nutrition for the whole population, despite the anticipated serious decrease in imports. To do this, food supplies had to be achieved by increasing home production of food and naval protection of imports; next, food had to be distributed equitably and effectively to all individuals in all parts of the UK. The actions the ministries took involved control of:

- all production, acquisition and distribution of food and farm requirements
- those imports that survived the journey
- all agricultural and factory production of foodstuffs, and
- distribution at all stages from port, farm or factory to individual.

It also required an increase in home production through an expansion in:

- land that could be used for producing food, and
- labour working on producing food.

Lightowler and Macbeth (2014) consider in greater detail than reviewed here all the controls from farm production or importation, further processing, dispersal, transportation, distribution, food rationing and foods off the ration; and importantly they discuss information available about the nutritional results. It should be noted that, at great loss of ships and men, the merchant navy, in convoys with the Royal Navy, still crossed the Atlantic carrying some supplies, but a great deal of food was sunk and the amount imported greatly reduced. Meanwhile, scarcity of space on precious ships stimulated new technologies for drying, preserving and packaging foods to last longer and to take up less space on ships (Ministry of Food 1946). Although most of us today would not choose to eat scrambled egg made from dried eggs, nor some other wartime products, these technologies have been converted to various uses in the processed foods of today.

Wartime Increase in Home Food Production and the Dissemination of Information

On the Farm

The Second World War followed a period between the wars of serious economic depression, and this included a depression in British farming, the result of which was that farmers and other landowners had fields and other land left uncultivated (Williams et al. 2012). Yet, in wartime conditions, this quickly provided the opportunity to bring more land into agricultural use. The War Cabinet minutes of October 1939 stated that 'the ploughing up campaign was making good progress' (National Archives 1939: 303). Although animals were still pastured on land unsuitable for growing crops, for example sheep in the mountains, where land could be cultivated for crops animal husbandry was reduced. Pastures for dairy cattle were the exception because the production of milk and milk products was encouraged for their nutritional value, and the consumption of liquid milk did increase. For other animals, however, the argument was that land to grow feed for animals competed with land to grow food for the population, and so fewer animals were raised; and throughout the war the consumption of meat by the population greatly reduced (Magee 1946). Magee's article about applying the science of nutrition discusses the nutritive value per acre of different vegetable crops, emphasising in particular the value of potatoes. Another example is that sugar beet crops were increased to reduce the need to import cane sugar.

As well as increasing the amount of land used for agriculture, especially cultivation, there was a need to increase the labour force on the farms, as the fit and able men had been called up for military service all over the world. Women entered occupations that they had not been in previously, and this included all aspects of agricultural work – manual labour on the land and driving machines, including the upkeep of those machines. The Women's Land Army had been formed in 1917 during the First World War but was disbanded in 1919. It was reformed in June 1939 before the outbreak of the Second World War, as war was already anticipated. Some two hundred thousand women volunteered to make agriculture their contribution to the war effort. Some continued in agriculture in peacetime up to 1950[2]. Furthermore, children worked on the farms in school holidays and there were working holiday camps for all the family on the farms at harvest time (Moore-Colyer 2004). Later in the war, German and Italian prisoners of war were employed in farm work, some of whom stayed in the UK after the war.

Off the Farm

For increased home production, apart from on the farms, again both land and labour had to be increased. Cities, towns and many villages already had fields divided into strips for individual families to grow food, called allotments. Under the government's campaign to 'Dig for Victory' (an echo of the similar campaign in the First World War), not only were allotments quickly booked up, more land put into allotments, and home gardens and lawns dug up for growing vegetables, but also every piece of public or otherwise accessible land that could be used was found and cultivated – beside railway lines and on stations, beside roads and lanes, in city and town parks, and more. In the Imperial War Museum there is a photo of a bomb crater filled with new soil in which rows of vegetables were being grown. With many men of fighting age away fighting, killed or prisoners of war, many women were working not just in the land army on farms but also replacing the men in factories of all sorts. So, the extra labour at home came from the elderly, the children after school, and above all from the adults out of their working hours. Thus, for those who had the opportunity, 'leisure' became the physical toil of growing food for the family. As a form of outdoor exercise, it also promoted health.

To describe the extra effort by everyone, including the destruction of lawns and flower beds in home gardens and public parks, as being just part of the war effort would be an appropriate but an insufficient explanation. Indeed, from conversations with those who were adults at the time, I learned that wartime conditions seem to have increased the ideals of community and neighbourly cooperation, and there was popular disapproval of those who tried to manipulate market conditions unfairly. Patriotism fostered by opposition to an enemy would have been part of this community feeling, as would details such as knowing that the food-rationing system affected all; even the royal family had ration books. The great efforts put into propaganda by the government would also have been important. The 'Dig for Victory' campaign encouraged people to grow vegetables for themselves and their families, with posters showing healthy, happy children. Public information in books, leaflets and on the radio gave information about growing vegetables (e.g. Middleton 1942) and recipes for cooking mostly with vegetables, with little meat or fat (e.g. Bruce 1942). Both these books were based on popular radio programmes. I also recall a page from my mother's handwritten notebook of recipes, written in my grandmother's handwriting for a 'butterless, eggless, milkless cake'. Whilst the food-rationing system was famously fair to all, with extras for pregnancy, early motherhood and certain medical conditions, or adjusted for infants and different childhood age groups or for preferences or beliefs such as vegetarianism, nevertheless the rations were limited to a calculated nutritional minimum. So, the limitation was also a stimulus to provide more food for oneself, for one's family and often to be shared with neighbours. In early 1945, I can remember our family receiving a gift of some fresh eggs which caused great pleasure. It particularly struck me, as I had just returned to the UK after an evacuee childhood in the USA unaware of any food shortages.

The aspects of greater production of food within the country and the dissemination of information about growing vegetables and about recipes appropriate for the wartime situation are themes to consider again in this chapter with a view to seeing if lessons can be learned from the 1940s that could be useful for supporting home food production, nutrition and sustainability in the twenty-first century. Hopefully, it will not be a matter of war at sea and the sinking of tons of food in merchant ships, but with contemporary uncertainties of Brexit and some nations' threats of protectionism, in 2019 much is unknown about future British and global trade. What will be the effects on food importation? The chances of inflation in the cost of food, in the UK at least, seem very high, which will affect the standard of nutrition of certain sectors of the population, whose food security is already at risk, and therefore their health (for an overview of food insecurity and health outcomes, see Gundersen and Ziliac 2015).

Allotments in Britain

Long before the legislation of 1908 and then 1922 and 1925, which required local authorities to provide allotments 'for all' at a very cheap rent, in England in rural areas there had historically been strips of land available for poorer members of a community to grow food for their families (Astill and Grant 1988). By the mid twentieth century, most English cities, towns and villages had areas of allotments, and for years these were well used by families in the locality. In both world wars, more had been created.[3] Today, in some cities there are waiting lists to get an allotment, but in other places the allotments have been discontinued and the fields built over for housing. However, a point relevant to this chapter is that in some of the remaining allotment fields there are still some empty allotments available for rent. With the cost and scarcity of land in the UK, and with the reduction in size of gardens associated with many of the recently built houses and apartments, and the increasing development of extensions or even new houses in previous home gardens, why is the land resource of allotments to rent not competed for everywhere?

In my Oxfordshire village, there is a field of allotments and I rent a half-allotment of 4 metres by 23 metres, where I grow vegetables and some fruit. In general, my area of the UK is affluent with little unemployment and, although many of the allotments are worked by retired folk like myself, there is a noticeable increase in plots worked by younger people whom one sees at weekends or on summer evenings. Some bring their children to make the work a family activity. Yet, every community has poorer members of society, maybe unemployed, but I can think of none of these among the allotment holders of our village, even though to rent a half-allotment only costs five pounds sterling per annum. In 2018 in our village, seven out of the sixty-four half-allotments remained available for rent. Why is there no competition for these? Will that change if fresh food prices rise? Our Allotments Association is not a closed allotment society and allotment availability is locally advertised. Whereas the allotments are within walking distance from any part of the village, some allotment holders come from further away. In other areas, especially in urban areas, availability may be limited, and/or access may require a journey of one or two buses, especially affecting those who might

Figure 7.1 South-east view across allotments in Kirtlington, Oxfordshire. Photograph by the author.

benefit most from the facility of growing their own food, but do not find access easy.

In 2015 and 2016 I interviewed allotment holders in our village and asked why they chose to work an allotment. Here is a small selection of answers that are representative of other answers:

'I know how the food is grown, with no use of pesticides'
'It's good to eat more vegetables'
'Our family is vegetarian'
'I like being out here; it's calming and quiet and healthy'
'It's good exercise'.

Almost all the people I interviewed gave answers that one way or another were linked to health. Just one person mentioned money. No one mentioned sustainability. Personally, I grow some vegetables in my small home garden and some in my rented allotment. In analysing my own views for doing this work, healthy food and exercise are significant factors, but there is also the convenience of picking the food when needed (and of course ready), rather than travelling to one of the nearby towns to a supermarket, whilst fresh food in the small village shop, without economies of scale, is understandably expensive. So, my views relate to health and exercise, but also to convenience of having food without travel. To all of this I should add the pleasure of a simple sense of achievement. I noticed long ago that my father claimed that the potatoes he had grown tasted better than any bought potatoes. I reckon that almost any grower of vegetables at home or on allotments finds that these vegetables, even if of less than perfect shape and colour, taste better than the beautifully shaped and coloured vegetables bought at a supermarket.

Figure 7.2 North-east view across allotments in Kirtlington, Oxfordshire. Photograph by the author.

This seems to parallel the study by Bratanova et al. (2015), which showed that subjects found that the food which they thought had been ethically grown (e.g. 'Fair Trade', 'organic' or 'locally produced') tasted better than the food in supermarkets of which they did not know the origin.

In summary, these reasons seem to me to be those of people who can afford options and choose to work their allotments. This does not explain why those who cannot afford options, but who live in a place where some allotments are available within walking distance, are not taking advantage of the allotment system. At least in our village we have a community effort three times a year when surplus foods from our allotments are donated and then 'sold' on a stall set up on our village green for whatever price the purchaser wishes to pay, and at the end of the morning what is left is given away. The resulting money is then split between different village charities, such as equipment for the small children's play area. All of this demonstrates community feeling but seems a long way from food sustainability.

Food Resourcing for Families by the Efforts of Individuals

In the Second World War, it was calculated that 1.3 million tonnes of food were grown by private individuals of all socio-economic levels on the allotments in Britain.[4] Hypothetically, there is a similar opportunity for food sustainability in the twenty-first century if people are willing to take on an allotment and if enough allotments are or become available for them to do so at an affordable rent. At this point, I want to mention a different kind of community garden system, where people cooperate on one piece of land to grow vegetables and fruit for the benefit of the whole group. An example of this can be found in Toxteth, a generally poor area in Liverpool, UK. Here there is a community garden where food is grown by communal effort, which serves not only to provide fresh food for consumption by the community but helps to address issues of social isolation. It was initiated as a community anchor for the neighbourhood by the community organisation Squash Nutrition. On Fridays anyone can come to the garden and buy a simple hot lunch made from the produce of the garden. Community garden food projects exist elsewhere, frequently fostered by a charity in conjunction with a local authority. Again, the work of growing the food is carried out by volunteers in their 'leisure' time, and some of the projects are designed to encourage work by people with special needs. One point remains: why do so few people take on the work at these projects to obtain fresh food affordably? One answer is, without doubt, that such projects are few and far between, and thus not within easy reach for most. Nevertheless, where accessible community gardens or allotments do exist, why do so few take part in the opportunities they offer?

There could be an interesting cross-disciplinary anthropological study, related to food and nutrition, gained from interviewing people in a diversity of socio-economic situations about whether or not they do, or would, take part in working either a personal allotment or joining others in work on a community garden. Of course, the research would need to look into whether such options are even available to them, and what would be needed to provide more such places. Whilst any such study in an affluent area might be interesting, it is needed far more urgently in other areas, for example in circumstances where in 2018 the food banks offer food to more and more people in real need (for more information on food banks and need see, for example, the Trussell Trust[5]). Therefore, the anthropological and cross-disciplinary research should be carried out in diverse socio-economic situations and cover urban, suburban and rural settings.

I return to the question of why, with allotment rents so cheap, there is no waiting list for plots where none are currently available; and why, in areas with available allotments in an accessible location, they remain unrented. Part of the reason in our village may be due to local affluence. In most families that affluence is gained by more than one working adult in the household, long hours at work, often continuing work at home and, for the majority in this village, time spent on commuting and other travel. Simply to find time for allotment work in such families might seem a luxury, whereas in wartime circumstances there had been both a necessity and the feeling of it being part of 'the war effort'. Those remaining at home in the UK during the war were also working long hours, and if commuting it would have taken extra time, either by bicycle or by limited public transport, as petrol was scarce and rationed. In 1945 my father left for work before our breakfast and returned home for 7 PM dinner, and then after dinner usually worked on papers despite all of us in a large family living our lives around him in the same room. He went to work on Saturday and Sunday mornings and, except for during any holidays, the only time to be with him was if we helped him in the vegetable garden on Saturday and Sunday afternoons.

However, not everyone is affluent nor works long hours. As there are people today who are unemployed or underemployed, retired or home carers, the anthropological study suggested should include how people's time is spent when not working. Perhaps no formal research is needed to verify that television and social media these days compete, like a drug, for our leisure time. One can walk down any street of houses and see the flickering light of television sets through front windows. It has been argued that for those in most financial need, television is the cheapest form of entertainment. Compared to going out for entertainment, this is true, but what is it about modern life that results in so much time being spent on passive entertainment? So, the anthropological question becomes what makes people choose so much passive entertainment over more active options. Is there a vicious circle of inactivity such that some of those who would most benefit from the exercise of working an allotment and the nutrition of fresh vegetables are more likely to spend hours of leisure time in front of the television, perhaps nibbling junk food? Of course, this is a caricature of a lifestyle, and it lacks, as far as I know, any serious research on the subject, beyond (ironically) what is portrayed on television. If appropriate research should identify this caricature as reality, the researchers might go on to hypothesise on what efforts might change the culture of those *un*employed or *under*employed television watchers, whose nutrition is poor and levels of physical activity unhealthily low. There is an interesting recent development in British television that provides programmes about saving money by growing one's own food and learning how to cook meals with more vegetables and less meat. I do not believe that these are government-financed, but they strike me as an echo of the wartime radio programmes, deliberately supported by the Ministry of Food, on similar topics (e.g. Bruce 1942; Middleton 1942).

Waste

Another aspect of lifestyle, not only in the war years but also through the depression in the United States as much as in Europe, was how every bit of waste should be reused. There were recipes for using up leftovers, for boiling up bones and other foods for stock, and for using bits of animal carcass that today may be thrown out. In the Second World War, even in individual families in urban areas, any scraps not suitable for some leftover recipe might be saved to feed a pig[6] or some chickens, or perhaps a pet. Such care with waste remains significant for many people around the world today, but not in many of our affluent societies, where there is food waste at every level. Indeed, people with a garden or allotment are more likely to put waste vegetable matter into a compost pile or bin, and all are now encouraged by local authorities to use municipal recycling bins for food waste. The collection of unsold vegetables from markets and supermarkets by charities has increased and is seen in a new waste-conscious perspective, such that large UK supermarkets make known their policies on what happens to 'just out-of-date' or mislabelled foods, such as being donated to charities. Some impoverished individuals search through rubbish skips and discarded food at the end of market days (as discussed by Black 2007).

Conclusion

In this twenty-first century, there are economic, political and climate-related threats to the standards of living of some sectors of many populations globally. These threats make it imperative to consider how sustainability of food can be achieved such that a healthy minimum nutrition is accessible to all. In a society, such as in UK, where health care, when needed, is supported nationally, then a logical corollary is that minimum healthy nutrition should also be of national importance. Insufficient access to fresh food already affects significant numbers of people, even in the so-called economically developed countries; charitable food banks have increased in number, and are needed, but are never sufficient. While economic inequalities seem inherent in an enterprise system, should not government support be directed at providing land opportunities, good transport possibilities and media encouragement for those most in need of fresh vegetables so that they could better access allotment options? However, if provided, would they do so?

In Britain in 2018, once again, a great proportion of food is imported. In the national statistics of 2015, only 52 percent of the food consumed in Britain was of UK origin, 29 percent came from other EU countries, 2 percent from non-EU European countries and the remainder from beyond Europe (Department for Environment, Food and Rural Affairs 2016).[7] A few generally ignored voices argue against 'food miles' because of the carbon emissions of carrying food great distances, but there are also carbon emissions in heating greenhouses in unsuitable climates. Surely, the fundamental problem lies in the modern culinary culture that one should be able to buy any food in any month of the year, even if it is locally out of season or totally inappropriate for the local climate all year round. In 2017, unusually severe winter weather conditions in southern Spain destroyed acres of vegetable crops and thus their export from Spain to the UK, where supermarket shelves briefly became bare of these vegetables, and retailers took steps to 'ration' what a purchaser might buy. This immediately shows one effect of the risk of depending on the importation of vegetables, even some that could be grown on a UK allotment, as well as others that are out of season in Britain. This argument might lead one also to think of the welfare of the populations in the vegetable-growing and exporting areas, and the sustainability effects on them should there be a fall in other areas importing their produce. Taking a global perspective, one should not avoid discussions of the food sustainability of the local populations where land for cash cropping has replaced subsistence farming, let alone the increasing pattern of land leasing by foreign nationals and foreign governments, precisely for the exportation of food or other commodities to their own countries. I leave it to be debated elsewhere whether global trade in vegetables does or does not benefit the nutrition of more people globally. With climate change, we are told that abnormally extreme weather conditions will become more common, and it could be argued that swift options for urgent food trade are essential for affected areas, whereas, in general, as much home production as possible should be encouraged. At the local level, it should not be forgotten that crop failures also exist in home gardens and on allotments, but patterns of communal exchange seem to be common between allotment holders.

So, with food imports again so high in Britain, and while we wonder what effects Brexit and protectionism might have on food prices, are there lessons for the objectives of food sustainability and a minimum nutrition for all that could be learned from the governmental actions in the Second World War? In contemporary peacetime society it is unlikely that such complete governmental control of so many aspects of daily living would be tolerated. Such control of all imports, control of the production of all farms and factories, and control of the wholesale and retail distribution, including ration books for all individuals, would be the antithesis of the market economy and free enterprise. Objection in so many quarters against some of the food standards adopted by the European Union, along with other measures such as fish quotas and farming regulations, well illustrates the dislike of such controls today.

It is the contention of this chapter that within our current political climate, only those aspects of the wartime government measures that encourage, rather than enforce, greater home food production on a voluntary basis would succeed. The question, therefore, becomes what would encourage voluntary actions in this regard. Presumably, rises in the price of foods might encourage voluntary action. Other possibilities are that stimulation for healthy physical activities and healthier nutrition might persuade some people, whilst for others it might be a concern for the environment and the threats of global warming. The Lancet caused considerable media attention by combining the challenges of global warming and healthier nutrition in an article (EAT-Lancet Commission 2019), which argued that red meat consumption should be greatly reduced (to 98 grams per person per week) and fruit and vegetables greatly increased. Worth noting is that, although measured differently, the meat consumption recommended in the article is even lower than that provided per week by UK's Second World War rations, and that the Lancet article was immediately followed by articles critical of its nutritional claims (e.g. Ede 2019; Harcombe 2019). Even with such public debate in the media, the unanswered question is whether a voluntary basis for growing more of one's own food would in peacetime ever become as common as in wartime, or would be undertaken by those who need it most. Enormous shifts in attitude would seem to be required ... but is anyone researching this?

Helen Macbeth
Oxford Brookes University, Oxford, UK

Notes

1. In particular, Lord Woolton and Jack Drummond.
2. For information on Women's Land Army in both World Wars, see https://www. womenslandarmy.co.uk. Accessed on 31 January 2019.
3. For information on allotments in UK, see https://www.allotment-garden.org/ allotment-information/allotment-history/. Accessed on 31 January 2019.
4. See note 3.
5. See https://www.trusselltrust.org/. Accessed on 31 January 2019.

6. Currently there is a UK and EU ban on feeding kitchen and restaurant waste to pigs and chickens. This ban is increasingly being questioned (e.g. zu Ermgassen et al. 2016) because of its value in recycling waste as feed.
7. I do not know to what extent, if any, home food production was included in this DEFRA calculation.

References

Astill, G. and Grant, A. (1988) *The Countryside of Medieval England*, Basil Blackwell, Oxford.

Bratanova, B., Vauclair, C.-M., Kervy, N., Schumann, S., Wood, R. and Klein, O. (2015) Savouring Morality: Moral Satisfaction Renders Food of Ethical Origin Subjectively Tastier, *Appetite*, 92, 137–49.

Black, R. (2007) Eating Garbage: Socially Marginal Food Provisioning Practices. In MacClancy, J., Henry, J. and Macbeth, H. *Consuming the Inedible: Neglected Dimensions of Food Choice*, Berghahn Books, Oxford: 141–150.

Bruce, P.J. (1942) *The Kitchen Front*, Nicholson and Watson, London.

Darke, S.J. (1979) A Nutrition Policy for Britain, *Journal of Human Nutrition*, 33, 438–444.

Department for Environment, Food and Rural Affairs (National Statistics) (2016) *Food Statistics Pocket Book 2016*, Crown Copyright, London, p. 22.

EAT-Lancet Commission (2019) *Food in The Anthropocene: the EAT-Lancet Commission on Healthy Diets From Sustainable Food Systems.* Published at: https://www.thelancet.com/commissions/EAT. Accessed on 27 January 2019.

Ede, G. (2019) EAT-Lancet's Plant-Based Planet: 10 Things You Need to Know, *Psychology Today* Blog. Published at: https://www.psychologytoday.com/us/blog/diagnosis-diet/201901/eat-lancets-plant-based-planet-10-things-you-need-know. Accessed on 31 January 2019.

Gundersen, C. and Ziliac, J.P. (2015) Food Insecurity and Health Outcomes, *Health Affairs*, 34(11): 1830–1839.

Harcombe, Z. (2019) The EAT Lancet Diet is Nutritionally Deficient. Published at: http://www.zoeharcombe.com/2019/01/the-eat-lancet-diet-is-nutritionally-deficient/. Accessed on: 31 January 2019.

Huxley, R.R., Lloyd, B.B., Goldacre, M. and Neil, H.A. (2000) Nutritional Research in World War Two: The Oxford Nutrition Survey and its Research 50 Years Later, *British Journal of Nutrition*, 84(2): 247–251.

Lightowler, H. and Macbeth, H. (2014) Nutrition, Food Rationing and Home Production in the UK during the Second World War. In Collinson, P. and Macbeth, H. (eds) *Food in Zones of Conflict: Cross-disciplinary Perspectives*, Berghahn Books, Oxford: 107–121.

Magee, H.E. (1946) Application of Nutrition to Public Health, *British Medical Journal*, 1(4447): 475–482.

Middleton, C.H. (1942) *Digging for Victory: Wartime Gardening with Mr. Middleton*, George Allen & Unwin, London.

Ministry of Food (1946) *How Britain was Fed in War Time: Food Control 1939–1945*, HMSO, London.

Mitchell, B.R. and Deane, P. (1962) *Abstract of British Historical Statistics*, Cambridge University Press, Cambridge.

Moore-Colyer, R.J. (2004) Kids in the Corn: School Harvest Camps and Farm Labour Supply in England 1940–1950, *The Agricultural History Review*, 52(2): 183–206.

National Archives (1939) Minutes of the War Cabinet, Catalogue Reference CAB/65/1/37: 303.

Sinclair, H.M. (1944) Wartime Nutrition in England as a Public Health Problem, *American Journal of Public Health*, 34, 828–832.

Theien, I. (2009) Food Rationing during World War Two: A Special Case of Sustainable Consumption? Published at: https://journals.openedition.org/aof/6383. Accessed on 2 February 2019.

Williams, C.A., Taylor, R. and Martin, J. (2012) *Wartime Farm*, Open University, Milton Keynes.

zu Ermgassen, E.K.J., Phalan, B., Green, R.E. and Balmford, A. (2016) Reducing the Land Use of EU Pork Production: Where There's Swill, There's a Way, *Food Policy*, 58: 35–48.

CHAPTER 8
FOOD AND SUSTAINABILITY IN THE TWENTY-FIRST CENTURY: HOW PLACES IN THE UK ARE WORKING TO MEET THIS CHALLENGE

....

Lucy Antal

Introduction

This chapter will consider the topic of food and sustainability in the twenty-first century in relation to the work of the Sustainable Food Cities Network[1] within the UK. There will be a specific focus on the six places that have received funding to host an officer, comparing and contrasting the different approaches taken, as well as examining the host organisations for each officer – which range from public health organisations to community non-governmental organisations (NGOs) to government departments.

Issues of food access, food surplus and waste, ethical procurement of food, food cultivation, food for health and food as an economic driver are important considerations for the Sustainable Food Cities Network (SFC). Building managed networks and partnerships has been key to the successes gained in each place. Each of these places faces significant challenges relating to poverty, citizen health, environment and the economy, and they plan to use sustainable food as a stimulus for change. This chapter will examine how this is being achieved, and the challenges faced.

Setting the Scene

To start at the beginning, what do we mean by the phrase 'sustainable food cities'? We, within the SFC, define 'sustainable' thus: *sustain* – to encourage and assist, to support physically or mentally, to continue, uphold and affirm;

and *able* – having the resources, opportunity and support to accomplish something. We want to work together to sustain healthy food activity so that people are able to live well, eat well and be well. As Virginia Woolf put it: 'One cannot think well, love well or sleep well, if one has not dined well' (Woolf 1929: 18).

Across the world, people are starting to recognise the role that food could play as a catalyst for changes in behaviour to solve social, economic and environmental problems (WRAP n.d.). As Beddington (2009: 1) put it:

> There is an intrinsic link between the challenge we face to ensure food security through the 21st century and other global issues, most notably climate change, population growth and the need to sustainably manage the world's rapidly growing demand for energy and water. It is predicted that by 2030 the world will need to produce 50 percent more food and energy, together with 30 percent more available fresh water, whilst mitigating and adapting to climate change. This threatens to create a 'perfect storm' of global events.

Whether related to diet-related ill health, food insecurity or food waste, or to intensive farming and industrialised food production, both of which impact on biodiversity and climate change, food not only contributes to some of our biggest dilemmas but can also form an essential part in the ways that we can solve them (Goodwin 2015). In the UK, the Sustainable Food Cities Network is interested in transforming food culture and the food system through positive action developed in cross-sector partnership. It is important to note that although the network is called the Sustainable Food Cities Network, it is a network of places rather than just cities. Members include towns, boroughs and wards.

The SFC approach involves developing a partnership of local public agencies, businesses, academic institutions and non-governmental organisations (including charities and communities), who work together with the ultimate goal of making healthy and sustainable food easily accessible for all residents within their area. The Sustainable Food Cities Network enables people and places to share challenges, explore practical solutions and develop best practice on key food issues. The SFC describes this approach as the 'Three Ps', by looking at how to drive positive change by:

- establishing an effective Food *Partnership*
- including healthy and sustainable food goals in *Policy*
- creating and helping to deliver a food strategy and Action *Plan*.[2]

Partnership

The success of any sustainable food programme relies on getting the right people involved in the way that works best for the place. One size does not fit all. This could be an informal steering group or a legally constituted body. The key strength of a food partnership is being able to bring stakeholders together to work on positive solutions. Egos need to be left at the door.

Policy

To create really significant improvements, food should be at the heart of policy. Whether the policies relate to procurement, health and well-being, economic development or urban planning, it is important to identify both the policies and the departments within local and national government, or more generally the whole public sector, that have relevance to food. The need is then to establish how to adapt the policies and the departments so that they enable sustainable food choices for all to become the norm. The policies should not be viewed as 'special' or particularly 'green', but something that is common practice and embedded into daily life. Taking a whole systems approach is also vital, recognising that food is a cross-sector signifier that impacts on a range of policy. For example, insufficient access to food (i.e. food insecurity) has a correlation to poor health indices, particularly in the North of England (Due North Report 2014). It is notable that food rarely receives an acknowledgement as a separate entity within political systems. There is no minister solely for food, nor do local councils appoint a cabinet member to oversee food. Instead it is shunted into health, or visitor economy, and thus is viewed solely through that prism, rather than the whole system approach that the Sustainable Food Cities Network advocates.

Plan

Having a plan is essential. The food partnership needs to develop an ambitious but achievable food strategy and an action plan for a programme that is based on the concept of an integrated place. Any town, city, borough, district or county can join the UK Sustainable Food Cities Network as long as it has a cross-sector food partnership working to create a better food system. The key to the cooperation of the network is to be willing to share successes (and failures!) and to be interested in learning from others.

Identified Challenges

Food Insecurity or Food Poverty

Both chronic and acute, a lack of food is having a big impact on the health, mental acuity and the resilience of citizens to be able to manage their lives. As

Dufour and colleagues write: 'Nutrition and resilience concepts are strongly interlinked: [good] nutrition is both an input to and an outcome of strengthened resilience. Reducing malnutrition is crucial to strengthening resilience because well-nourished individuals are healthier, can work harder and have greater physical reserves; households that are nutrition secure are thus better able to withstand external shocks' (Dufour et al. 2014: 1).

This is not just about food access, but also about low wages, transport options, working life shift patterns and changes made to state support tax benefits – for example, the UK's so called 'second bedroom tax' whereby housing benefit is cut if you live in a council or housing association home and are classed as having a spare bedroom. In the UK there have been trials of a new system of Universal Credit, designed to combine various tax relief schemes and benefit payments into one system of calculation (e.g. Cash 2016; Martin 2016; Hickman et al. 2017; Omar et al. 2017). Research conducted by Liverpool City Council's benefits team demonstrated that the cumulative effect of changes in benefit payments since 2010 had negatively impacted upon some 50,000 households. The report (Liverpool City Council 2017) – which uses the Government's impact analysis reports (Department for Work and Pensions 2012) and the data held by the council – shows that the changes to working age benefits since 2010 have affected around 55,000 households (one in four) – with the long-term sick and disabled, children and women being disproportionately hit.

Additional research by the housing charity Shelter, carried out in conjunction with YouGov, found 44 percent of low-paid renters had been forced to cut back on basic items, including food, clothes and toys for their children, in order to pay for their home (Shelter 2017).

Healthy Lives

The correlation between health and food is fundamental to the sciences of nutrition and dietetics (e.g. Tapsell 2016), and their link with variables such as age and economic status is regularly researched and described in medical, nutritional and dietetic journals (recent examples are: Williamson et al. 2017; Burgoine et al. 2018; Spence et al. 2018). In particular, statistics published by the Malnutrition Taskforce revealed that 30 percent of elderly people admitted to hospital were malnourished; and malnourished patients will stay between five and ten days longer in hospital (Wilson 2013). This in turn has an impact on health care budgets. Could a more tailored approach to nutrition help to reduce hospital admissions? Researchers are also examining the effect of food on mental health. A recent global study, 'Food Insecurity and Mental Health Status', concluded that chronic food insecurity is enough to bring about long-term changes to mental health (Jones 2017).

Food Availability

We live in a world of increasing on-demand supply for anything at any time. But not everyone benefits (Walker et al. 2010). A move towards the centralisation of food purchasing into the 'one-stop' supermarket has led to a proliferation of large hypermarkets within urban locations, encouraging consumers to do all their food (and other household purchases) shopping in one space. This in turn has reduced the availability of food within local spaces, corner shops have declined, with fast food outlets and betting shops taking their place (Larsen and Gilliland 2008). Not everyone can access the large stores if transport options are limited or if individuals do not have the spare income to pay for delivery (usually based on a minimum spend) (King et al. 2015). An article in the *Guardian* (Tickle 2016) highlighted the barriers faced by a pensioner in Bristol, a city with eighty-four supermarkets yet containing large estates without any food shopping provision, often referred to as 'food deserts' (see also Garthwaite 2016).

Food Waste or Surplus

There is a conundrum. Overproduction and determination to meet every possible consumer demand is leading to a food surplus that in turn becomes food waste (Aschemann-Witzel et al. 2015). At the same time, we have working family householders accessing food banks because they do not have the income capacity to manage an adequate weekly food budget. The Waste Resource Action Plan (WRAP) has identified that 4.2 million tonnes of food and drink are wasted each year, and could be avoided (Quested et al. 2013).

Diversity of Supply

The economics of food can be overlooked by policy makers outside of tourism (Sims 2009). The public sector has a role to play in supporting local and smaller suppliers, enabling a more circular economic approach. The Global Goals for Sustainable Development includes: 'Ensure sustainable consumption and production patterns' (MacArthur 2015). This can be applied particularly to food, reducing carbon footprint and food waste, and ensuring a more equitable distribution of assets and cash within a local economy context. The UK government is keen to promote use of local suppliers for public sector food procurement via the Government Buying Scorecard (Department for Environment, Food and Rural Affairs 2014).

Loss of Knowledge

Ignorance of food, where it comes from, how to grow it, how to prepare it and when and what to eat for optimum health benefits could be seen as the main contributing factors for all of the identified challenges listed above. In relation to this, a British Nutrition Survey of school children identified high levels of ignorance about food (British Nutrition Foundation 2017; Jamie Oliver Food Foundation 2017). Re-education, installing confidence in people's abilities to prepare and cook fresh food (Tull 2015), and explaining basic health messages are part of the 'whole systems' approach that the Sustainable Food Cities Network is championing.

Six Places

For the purposes of this chapter, the focus will be on the six places, all large towns or cities, awarded additional funding to employ a coordinator to facilitate this process. They are:

Belfast: Capital and largest city in Northern Ireland, Belfast has had a chequered recent past with 'the Troubles' but has undergone an economic and regenerative transformation following the Good Friday Agreement. It is a port city with a traditional shipbuilding industry (known for building the Titanic), with a population today of approximately 333,000 (http://www.belfastcity.gov.uk/council/Yourcouncil/yourcouncil.aspx).

Bournemouth and Poole: Bournemouth is a large seaside resort town on the south coast of England, while its neighbour Poole is a large coastal town and seaport. Together they are part of the South East Dorset conurbation and as such have a combined population of over 465,000 (Office for National Statistics, 2011 Census).

Cardiff: Capital and largest city in Wales, Cardiff is the chief commercial city for the country, and site of the Welsh Assembly. It is a significant tourist destination and seaport, with a population of approximately 346,000 (Office for National Statistics, 2011 Census).

Liverpool: As a port and city in North West England, Liverpool was previously very prosperous, but had a period of decline following the closure of several industries and the reduction in port traffic. However, it reinvented itself as a destination for tourism, and is known for its football and its music. The population approximates 473,000, and it is relevant to note that it is home to the oldest Black African community in Britain and the oldest Chinese community in Europe (Office for National Statistics, 2011 Census).

Newcastle: This city and port in North East England lies on the banks of the River Tyne. Previously coal mining and shipbuilding were its main industries, but now there is a concentration of cultural, business and knowledge activities. The population approximates 280,000 (Elledge 2015).

Stockport: This large town is part of Greater Manchester in North West England, on the banks of the River Mersey. It was previously a market town that developed into a textile industry town. It has a population of approximately 138,000 (Office for National Statistics, 2011 Census).

Each of these places was given enough funding to cover the post of a coordinator for thirty months and a small budget to facilitate the building of a strong food partnership and network within its region. It has been interesting to compare and contrast the different approaches to this task, and to examine whether the host organisation for each officer has helped, hindered or otherwise influenced the priorities of the programme.

The Focus on Themes

The Sustainable Food Cities Network focuses on six key food issues, which are, in no particular order of priority:

Promoting Healthy and Sustainable Food to the Public

Increasing public knowledge of, and interest in, healthy and sustainable food is the first step to creating an affirmative food culture. Suggested options include:

- Healthy eating campaigns – e.g. Breastfeeding and Sugar Smart Cities,[3] which aim to change public behaviour, particularly among hard-to-reach target audiences.
- Sustainable food campaigns to promote more public consumption of fresh, seasonal, local, organic, sustainably sourced fish, high animal welfare, meat-free and/or Fair Trade food.
- Developing a food charter that offers a food vision for the city or town, and encourages organisations to take action to turn that vision into reality.
- Creating opportunities for people to buy affordable, healthy and sustainable food through markets and mobile pop-up shops and restaurants, particularly in areas with no existing provision. Access to fresh food is key to achieving better health outcomes.

Tackling Food Poverty, Diet-Related Ill Health and Access to Healthy Food

In our contemporary society, food-related inequalities are creating increasing problems of hunger, obesity, malnutrition and diet-related ill health. Key options are:

- To establish a multi-agency partnership that includes key public and voluntary organisations. This partnership should assess and tackle the full range of issues that contribute to chronic food insecurity and acute food poverty in a joined-up and strategic way.
- To promote the Living Wage[4] through local authority commitments and/or via campaigns to raise employer awareness of the economic and social value impact of paying higher wages – and the consequences of not doing so.
- To increase the understanding of food insecurity issues amongst health professionals, welfare advisers, housing and voluntary organisations, in particular how food poverty can have a cumulative impact on mental and physical health outcomes.
- To increase the availability of healthy options in supermarkets, convenience stores, takeaways, vending machines and in catering settings such as nurseries, schools, hospitals, care homes and workplaces. The ambition is that healthy food should become the norm for all, not just for the privileged and educated.

Building Community Food Knowledge, Skills, Resources and Projects

Positive food change often comes from the energy, innovation and action of grass-root communities. Individuals and communities often have the knowledge, skills, resources and projects to make that change happen. The ways to support such positive food change include:

- Establishing a network for community food activists in order to help them share ideas and resources, co-create projects, and to direct them to advice, training, grants and toolkits.
- Mapping and making available assets such as green- and brownfield sites, or redundant buildings that can be used for community food-growing and other projects.
- Lobbying for the incorporation of urban food-growing possibilities in local development projects through the creation of roof gardens and community growing spaces in residential housing and commercial developments.

- Increasing community food growing through additional provision of allotments, edible landscaping (Creasy 2009), and through initiatives such as The Big Dig[5] and Incredible Edible.[6]

Some of the above would be very hard to achieve without changes to local authority policies and practice, aimed at enabling individuals and communities to get improved access to resources that could be used for food enterprises or projects. This is why it is so necessary to approach the problems by building a strong cross-sector food partnership.

Promoting a Vibrant and Diverse Sustainable Food Economy

To transform food cultures and the systems that support them, food should be good for local economies, businesses and jobs, as well as for the people, their cultural acceptance and the planet. This requires:

- Developing strategies, policies and services that support the development and long-term success of healthy and sustainable food businesses.
- Asking for support for new sustainable food entrepreneurs via vocational training and business planning, finance, development advice, support and/or grants.
- Helping producers to improve how they connect with local consumers and/or achieve better access to wholesale and retail markets through events and marketing initiatives.

Transforming Catering and Food Procurement

Catering and procurement offers one of the most effective ways to drive large-scale changes in healthy and sustainable food. A moral argument also occurs for the public sector, such as hospitals, that they should consider the health and social value of their procurement policies. DEFRA's recent Health and Harmony procurement consultations have advocated 'public money for public good'. This can be achieved via the following:

- Adopting a place-wide Sustainable Food Procurement policy that incorporates specific commitments on a range of health, sustainability and/or ethical issues.
- Persuading individual public sector organisations to adopt specific food policies – e.g. healthier catering and vending facilities, 'tap water only' and ethical standards such as cage-free and Fair Trade foods.

- Encouraging public and private caterers to achieve accreditation on quality, health, sustainability and/or ethical food criteria, such as the Food for Life Catering Mark,[7] Sustainable Fish Cities[8] and Good Egg Award.[9]

Reducing Waste and the Ecological Footprint of the Food System

Reducing food waste potentially has huge social, economic and environmental benefits. To manage the ecological footprint of waste, it is necessary to consider how food is produced, processed, packaged and transported, and the waste that occurs at each stage. For this it would be beneficial to:

- Establish a food-waste collection scheme for homes and businesses, and redirect this waste for composting, energy recovery (Anaerobic Digestion) or animal feed.
- Promote home and community food composting through awareness and education campaigns, and through the provision of composting tools, demonstrations, materials and sites for communities to use.
- Collect harvest surplus and unwanted produce from local farms and food-growing sites through a crop-gleaning/abundance and volunteering scheme to redirect harvested produce that is unwanted by retailers (e.g. The Gleaning Network).
- Collect and redistribute consumable surplus food to organisations that provide food for people in need (e.g. FareShare),[10] while working to raise the nutritional standards of the food being offered.

Challenges, Successes and Different Approaches

One of the main advantages of being part of the UK Sustainable Food Cities Network is that despite the different sizes of the places involved, the food partnerships are all dealing with similar issues. So, the network offers the opportunity to discuss the challenges and successes, and to share the ways in which the different partnerships are approaching them. A short survey was sent out to the coordinators managing the food partnerships in the six food cities/towns. A brief analysis of the responses to some of the questions follows, and demonstrates that there was diversity in answers, but some overlap. It is likely that there was far more commonality than demonstrated, and that diversity arose more from choosing what to put in a short response than from huge differences between sites. In discussing the responses, the city or town of each coordinator is used to refer to the coordinator responding.

Survey Questions and Summary of Answers

What did you/your organisation consider to be the biggest/most significant challenge to your city (place) at the start of the project? (e.g. food poverty / lack of interest in food amongst policy makers).

Several themes arose in the responses. Stockport and Cardiff identified that it was 'all new' and so an initial challenge was where to start. Even the meaning of the words 'sustainable food' had to be reviewed and a concept reached. Several sites mentioned that it had not been previously on any political agenda, and Newcastle found it hard to engage with the local authority in a consistent manner, whereas Liverpool found that the only sectors that recognised the importance of the topic were the public health sector and the third sector (National Audit Office 2010). Belfast had found it easier to engage with others because food poverty was well recognised in the city, and their response focused on details, such as no access to fresh food. Stockport specified the effects of recent policies of austerity. Generally, the sites were all concerned with strategies on how to bring it all together, to make the topic relevant and accessible in all sectors; Cardiff, however, identified an obstacle caused by a certain amount of protection of individual roles by some organisations. Newcastle more optimistically referred to the recognition of the benefits.

What was the biggest issue for each city? Any commonalities?

Using a systems context diagram (Figure 8.1), Food Newcastle produced a very eloquent response to demonstrate the complexity of cross-partnership working. This, alongside the problem of food and health inequality, came out as the two main issues for all six sites.

What has been the easiest thing to tackle/approach/make a difference to?

Again, Belfast referred to the issue of food poverty, whereas other sites referred to the building of new relationships to influence projects. Stockport referred to new links at neighbourhood level, Bournemouth had found working with businesses beneficial, and Liverpool had had success in building a strong network within the city that included the support of academic centres in the universities. Cardiff specifically mentioned working with the Fish Cities Campaign.

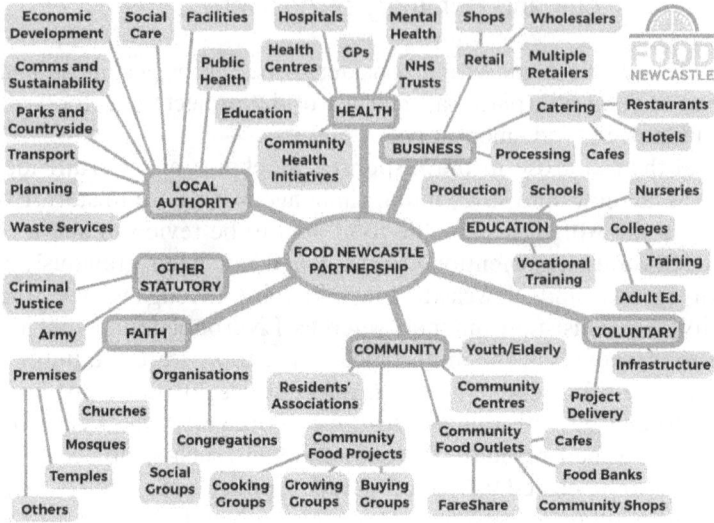

Figure 8.1 An example from Food Newcastle of how many potential stakeholders are involved in a food partnership. Courtesy Food Newcastle.

Conclusion

Building a sustainable food system takes time. After twenty-eight months engagement, all six sites are making progress but that progress is much slower than perhaps the funders had anticipated. Notable achievements include the following: Cardiff and Belfast achieved Bronze Sustainable Food Cities Awards; Newcastle established the first community food market in the SFC network, which now runs every third Saturday in the month; Stockport has taken on a 3.5-acre site to create a productive community garden, and levered £250,000 of support into it; Liverpool has influenced changes to the catering within the local hospital, and has one of the largest holiday hunger programmes in the UK; and Bournemouth has engaged with their food businesses via an award system. The coordinators of all the places cited agree that additional funding support and purchases from strategic bodies is needed. The work has only just begun.

Lucy Antal
The Food Domain, Liverpool

Notes

1. Sustainable Food Cities Network, http://sustainablefoodcities.org/about, last accessed 14 February 2019.
2. www.sustainablefoodcities.org, last accessed 14 February 2019.
3. Sustain and Jamie Oliver Foundation Sugar Smart website 2017 (www. sugarsmartuk.org, last accessed 14 February 2019).
4. Living Wage Foundation, Citizens UK initiative (http://www.livingwage.org.uk/ what-is-the-living-wage, last accessed 14 February 2019).
5. Sustain, The Big Dig project 2018 (http://www.bigdig.org.uk/, last accessed 14 February 2019).
6. Incredible Edible Network, Growing Together Partnership (http:// incredibleediblenetwork.org.uk/, last accessed 14 February 2019).
7. Soil Association, Food for Life Awards (https://www.soilassociation.org/ certification/catering/, last accessed 14 February 2019).
8. Sustainable Fish Cities Campaign (https://www.sustainweb.org/sustainablefishcity/, last accessed 14 February 2019).
9. Compassion in World Farming, Good Egg Award (https://www. compassioninfoodbusiness.com/awards/good-egg-award/, last accessed 14 February 2019).
10. FareShare (http://www.fareshare.org.uk/, last accessed 14 February 2019).

References

Aschemann-Witzel, J., de Hooge, I., Amani, P., Bech-Larsen, T. and Oostindjer, M. (2015) Consumer-Related Food Waste: Causes and Potential for Action, *Sustainability*, 7(6): 6457–6477.

Beddington, J. (2009) *Food, Energy, Water and the Climate: A Perfect Storm of Global Events?* UK Government Office for Science, London. Published at https://www. scribd.com/document/127062061/food-energy-water-and-the-climate-a-perfect-storm-of-global-events. Accessed on 21 January 2018.

British Nutrition Foundation press release (2017). Published at: https://www.nutri-tion.org.uk/nutritioninthenews/pressreleases/1059-bnfhew2017.html Accessed on 9 March 2018.

Burgoine, T., Sarkar, C., Webster, C.J. and Monsivais, P. (2018) Examining the Interaction of Fast-Food Outlet Exposure and Income on Diet and Obesity: Evidence from 1,361 UK Biobank Participants, *The International Journal of Behavioral Nutrition and Physical Activity*, 15(1): 71.

Cash, D. (2016) The Universal Credit Rating Group: Measuring Debt Ethically, *World Economics*, 17(4): 67–74.

Compassion in World Farming, Good Egg Award. Published at: https://www.compas-sioninfoodbusiness.com/awards/good-egg-award/. Accessed on 21 January 2018.

Creasy, R. (2009) *Edible Landscaping Basics*. Published at: http://www.rosalindcreasy. com/edible-landscaping-basics/. Accessed on 21 January 2018.

Department for Environment, Food and Rural Affairs (DEFRA) (2014) *A Plan for Public Procurement: Food & Catering Balanced Scorecard for Public Food*

Procurement, UK Government paper. Published at: https://www.gov.uk/govern-ment/publications/a-plan-for-public-procurement-food-and-catering-the-balanced-scorecard. Accessed on 21 January 2018.

Department for Work and Pensions (DWP) (2012) *Universal Credit Impact Assessment*, UK Government paper. Published at: https://assets.publishing.service. gov.uk/government/uploads/system/uploads/attachment_data/file/220177/univer-sal-credit-wr2011-ia.pdf. Accessed on 21 January 2018.

Due North Report (2014) *The Report of the Inquiry on Health Equity for the North 2014*, chaired by Margaret Whitehead, report prepared by the Inquiry Panel on Health Equity for the North of England, University of Liverpool and Centre for Local Economic Strategies.

Dufour, C., Kauffman, D. and Marsland, N. (2014) Enhancing the Links between Nutrition and Resilience. In Fan, S., Pandya-Lorch, R. and Yosef, S., *Resilience for Food and Nutrition Security*, International Food Policy Research Institute, Washington, DC, pp. 107–117.

Elledge, J. (2015) *Where Are the Largest Cities in Britain?* Published at: http://worldpopulationreview.com/countries/united-kingdom-population/cities/. Accessed on 9 March 2018.

Garthwaite, K. (2016) *Hunger Pains: Life inside Foodbank Britain*, Policy Press, Bristol.

Goodwin, E. (2015) The Rising Resource Challenge through the Eyes of Ines, WRAP. Published at http://www.wrap.org.uk/content/rising-resource-challenge-through-eyes-ines. Accessed on 21 January 2018.

Hickman, P., Kemp, P.A., Reeve, K. and Wilson, I. (2017) The Impact of the Direct Payment of Housing Benefit: Evidence from Great Britain, *Housing Studies*, 32(8): 1105–1126.

Jamie Oliver Food Foundation (2017) *A Report on the Food Education Learning Landscape,* AKO Foundation. Published at: https://www.schoolfoodmatters.org/ sites/default/files/%ef%80%a1%ef%80%a1FELL%20REPORT%20FINAL. pdf#overlay-context=news/food-education-learning-landscape-report. Accessed on 20 January 2019.

Jones, A.D. (2017) Food Insecurity and Mental Health Status: A Global Analysis of 149 Countries, *American Journal of Preventative Medicine*, 52(2): 264–273.

King, G., Lee-Woolf, C., Kivinen, E., Hrabovszki, G. and Fell, D. (Brook Lyndhurst) (2015) *Understanding Food in the Context of Poverty, Economic Insecurity and Social Exclusion*. Brook Lyndhurst for the Food Standards Agency in Northern Ireland, Crown Copyright.

Larsen, K. and Gilliland, J. (2008) Mapping the Evolution of 'Food Deserts' in a Canadian City: Supermarket Accessibility in London, Ontario, 1961–2005, *International Journal of Health Geographics*, 7: 16.

Liverpool City Council (2017) *Welfare Reform Cumulative Impact Analysis 2016: Interim Report 2017*, Liverpool City Council, Liverpool.

Living Wage Foundation, Citizens UK Initiative. Published at: http://www.livingwage. org.uk/what-is-the-living-wage. Accessed on 21 January 2018.

MacArthur, E. (2015) How the Circular Economy Can Help Us Achieve the Global Goals, *World Economic Forum*. Published at: https://www.weforum.org/ agenda/2015/10/how-the-circular-economy-can-help-us-achieve-the-global-goals/. Accessed on 9 March 2018.

Martin, J. (2016) Universal Credit to Basic Income: A Politically Feasible Transition?, *Basic Income Studies*, 11(2): 97–131.

National Audit Office (2010) *What Are Third Sector Organisations and their Benefits for Commissioners?* Published at: https://www.nao.org.uk/successful-commissioning/introduction/what-are-civil-society-organisations-and-their-benefits-for-commissioners/. Accessed on 22 April 2018.

Office for National Statistics (2011) Census. Published at: https://www.ons.gov.uk/census/2011census. Accessed on 22 April 2018.

Omar, A., Weerakkody, V. and Sivarajah, U. (2017) Digitally Enabled Service Transformation in UK Public Sector: A Case Analysis of Universal Credit, *International Journal of Information Management*, 37(4): 350–356.

Quested, T., Ingle, R. and Parry, A. (2013) *Household Food and Drink Waste in the UK 2012*, Final Report, WRAP, Banbury.

Shelter (2017) *Shut Out: The Barriers Low-Income Households Face in Private Renting*, Shelter.org.uk. Published at: https://england.shelter.org.uk/__data/assets/pdf_file/0004/1391701/2017_06_-_Shut_out_the_barriers_low_income_households_face_in_pivate_renting.pdf. Accessed on 21 January 2018.

Sims, R. (2009) Food, Place and Authenticity: Local Food and the Sustainable Tourism Experience, *Journal of Sustainable Tourism*, 17(3): 321–336.

Spence, A.C., Campbell, K.J., Lioret, S. and McNaughton, S.A. (2018) Early Childhood Vegetables, Fruit and Discretionary Food Intakes Do Not Meet Dietary Guidelines, but Do Show Socioeconomic Differences and Tracking over Time, *Journal of the Academy of Nutrition and Dietetics*, 118(9): 1634–1643.

Tapsell, L.C. (2016) Examining the Relationship between Food, Diet and Health, *Nutrition and Dietetics*, 73: 121–124.

Tickle, L. (2016) Five a Day? It's None a day in Britain's Urban Food Deserts, *Guardian*, 5 July 2016. Published at https://www.theguardian.com/sustainable-business/2016/jul/05/five-a-day-cities-shortage-affordable-fruit-veg. Accessed on 21 January 2018.

Tull, A. (2015) Why Teach (Young) People to Cook? A Critical Analysis of Education and Policy in Transition. Doctoral thesis, City University, London.

Walker, R.E., Keane, C.R. and Burke, J.G. (2010) Disparities and Access to Healthy Food in the United States: A Review of Food Deserts Literature, *Health and Place*, 16(5): 876–884.

Williamson, S., McGregor-Shenton, M., Brumble, B., Wright, B. and Pettinger, C. (2017) Deprivation and Healthy Food Access, Cost and Availability: A Cross-sectional Study, *Journal of Human Nutrition and Dietetics*, 30(6): 791–799.

Wilson, L. (2013) *A Review and Summary of the Impact of Malnutrition in Older People and the Reported Costs and Benefits of Interventions*, Malnutrition Task Force, UK. Published at: http://www.malnutritiontaskforce.org.uk/wp-content/uploads/2014/11/A-review-and-summary-of-the-impact-of-malnutrition-in-older-people-and-the-reported-costs-and-benefits-of-interventions.pdf. Accessed on 9 March 2018.

Woolf, V. ([1929] 1957) *A Room of One's Own*, Harcourt, Brace & World, San Diego, CA.

WRAP (n.d.) *Food Futures: From Business as Usual to Business Unusual*, WRAP report. Published at: http://www.wrap.org.uk/sites/files/wrap/Food_Futures_%20report_0.pdf. Accessed on 9 March 2018.

CHAPTER 9

FOOD AND THE PROBLEM OF UNCERTAINTY – REFUGEES AND THE SENSE OF SUSTAINABILITY: THE CASE OF KAREN FARMERS RETURNING TO THEIR VILLAGES FROM REFUGEE CAMPS ALONG THE THAI–BURMESE BORDER

Peter Kaiser

Introduction

According to the Food and Agriculture Organization (FAO),

> Sustainable diets are those diets with low environmental impacts, which contribute to food and nutrition security and to healthy life for present and future generations. Sustainable diets are protective and respectful of biodiversity and ecosystems, culturally acceptable, accessible, economically fair and affordable; nutritionally adequate, safe and healthy; while optimising natural and human resources. (Burlingame and Dernini 2010: 7)

Are we allowed to claim these prerequisites for food production and all nutrition-related behaviour? It seems crucial to highlight them at a time when humanity can damage the world we are living in for generations to come, and while we still have time, hopefully, to undertake important steps to prevent that kind of development. Nevertheless, not all people around the world are in a situation in which they can be concerned about saving the earth, because they are fully concerned with just surviving. The protection of the diversity of both plants and animals, and the welfare of farmed and wild species, as well as the avoidance of damaging or wasting natural resources or contributing to climate change may be a nice thought, but survival and subsistence are the most essential things for many people. Even further out of reach are the provision of good quality, safe and healthy food, social benefits, and educational opportunities. It can only be wishful thinking that they would also contribute to thriving local economies and sustainable livelihoods – both in producer countries and, in the case of imported products, in consumer countries.

Members of the Karen ethnic group have been interned in long-term refugee camps along the Thai–Burmese border since the mid-1980s. They have been both refugees[1] and internally displaced persons (IDPs).[2] Now, after the officially announced armistice between the Karen National Union (KNU) and the Burmese government in January 2012, the refugees have to return to their abandoned villages. Up to now, in the refugee camps, the former farmers have had no opportunity to pass on their knowledge of cultivation to their offspring, nor have they received any alternative training. Such training should be designed specifically to meet the needs of smallholder farmers and producers, with a special focus on women – such as training for small-scale income-generating new (cash) crops production or other skills so that they become more independent (for example; sewing with sewing machines). So, for those who return, some relevant knowledge of cultivation will be available to them. This chapter is about their situation and the problems to be faced both nutritionally and psychologically with their return to their previous settlements. The example of the Karen shows how the well-meant aid of refugee camps can lead to a loss of cultural knowledge, as well as of self-reliance and self-sufficiency.

Many Karen living in refugee camps have been victims of forced migration[3] which can be found across the Karen State and in other areas of the eastern part of Myanmar. Another type of forced migration is by military occupation-induced and development-induced displacement. This is generally caused by (a) confiscation of land – following armed conflict – by the (here, Burmese) army or other armed groups, including for natural resource extraction and infrastructure construction, or (b) predatory taxation, forced labour or other abuses. Myanmar[4] consists of fourteen provinces – or seven states, representing the areas of seven main ethnic groups, and seven divisions. All seven states are more or less mountainous, whilst the divisions are mainly plain areas, with the exception of Sagaing, Bago and Thaninthayi divisions. The border area comprises the Kachin, Shan, Kayah, Kayin, Mon and Thaninthayi provinces. All the border states and divisions are affected by militarisation and/or 'development'-induced displacement, as are a number of urban areas, including in the context of developing tourism and 'urban renewal' (South 2007: 16). Furthermore, there is livelihood-vulnerability-induced displacement: this is the primary form of internal and external migration within and outside of Myanmar. The main causes of this are inappropriate government policies and practices, limited availability of productive land and poor access to markets, which all lead to food insecurity and a lack of education and health services (South 2008).

Karen: Their Historical and Social Background

The first Karen-speaking groups are likely to have arrived in the area of today's Myanmar towards the end of the first millennium (Renard 1979). The

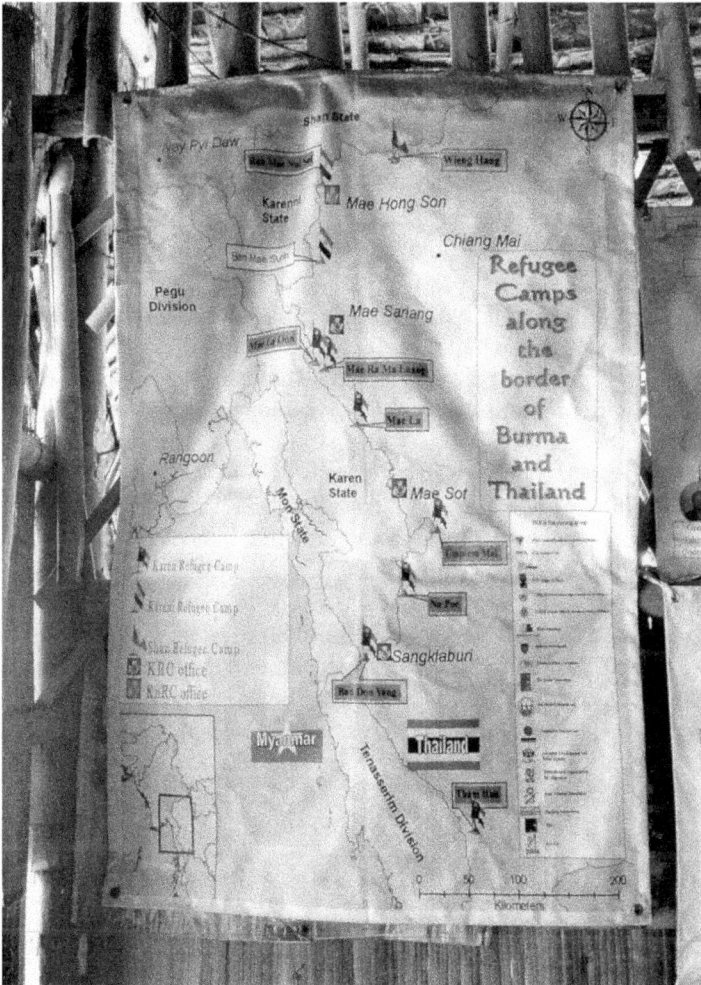

Figure 9.1 Map of the refugee camps along the Thai–Burmese border, displayed on the wall of a hospital at Mae Ra Ma Luang, 2017. Photograph by the author.

Karen make up approximately 7 percent of the total Burmese population, with a population of five million people (another four million live abroad).

The Karen are the third biggest ethnic group in Myanmar, after the Bamars (Burmese) and the Shans. However, the term 'Karen' refers to a heterogeneous number of ethnic groups that do not share a common language, culture, religion or material characteristics. A pan-Karen ethnic identity is a relatively modern creation, established in the 1800s with the conversion of some Karens to Christianity, and shaped by various British colonial policies and practices.

Recent History

The decades-long conflict between Myanmar's military, with its ethnic cleansing campaign, and the country's ethnic armed groups, such as the Karen National Liberation Army (KNLA) and others, has created one of the most protracted refugee situations in the world. First established in 1984, it is estimated that more than one hundred thousand Myanmar refugees currently live in the nine camps along the Thailand–Myanmar border. The vast majority of the refugees come from five states or regions in south-east Myanmar, close to or on the border with Thailand: Kayin (also known as Karen) State, Kayah (Karenni) State, Taninthayi (Tenasserim) Region, Bago (Pegu) Region and Mon State. Eighty percent of the camps' residents in 2014 were ethnic Karen, and many of the refugees were born in exile (The Thailand Burma Border Consortium[5] 2004, 2008a; The Border Consortium 2014). As the Thai authorities have prohibited refugees from travelling, farming or collecting firewood outside the camps, the refugees' coping mechanisms have been severely eroded. Since 2005, more than 83,000 people who fled from Myanmar to Thailand have resettled in third countries, with most going to the United States. These were predominantly skilled people – the so-called brain drain. Thailand ended the registration of refugees in 2006 and has maintained that those who are not documented are ineligible for moving to a third country. An estimated one million Burmese reside in Thailand, most of them undocumented migrant workers. Refugees face restrictions on movement and have limited opportunities for formal employment or for further education (Burma Link; Human Rights Watch 2012; UNHCR 2015a, 2015b).

Nutrition and Food

Traditionally, the Karen people have depended for their livelihoods on rice farming, forest products and shifting cultivation on hillsides (Marshall 1922 [1997]). Their main activities included:

- Cultivation of indigenous rice varieties as a staple food.
- Gathering of forest products (mushrooms from the forest and edible wild vegetables).
- Raising livestock such as chickens, pigs and cows.
- Cultivation of sesame and vegetables including cucumbers, pumpkins, bamboo shoots and eggplants.
- Cultivating tobacco to be used for smoking as well as for keeping insects away from rice fields.
- Where fishing is an option, they would fish (and include in their diet a very famous dish among the Karen known as 'nya u', or in Burmese 'ngape' – a strong-tasting dish of fermented fish).

Nutrition and Food in Transition

According to one man who was forced to leave his village and seek temporary shelter in the jungle three times in 1996:

> Each time we had no food. In the forest, relationships varied. Some shared food with others then left to look for roots together; others did not. I saw one family close to utter starvation, the two small children crying from hunger. The mother pitifully fed them roots which hadn't been boiled long enough; she probably didn't know what else to do. They had absolutely no possessions whatsoever, other than one pot, a machete and a small blanket. (Asian Human Rights Commission 1999: 24)

The living conditions during the flight differed completely from the situation at times of peace, as well as from the conditions in the refugee camps along the border. During the flight and migration, cultivating rice is not possible; consumption depends on what is carried, with only limited fishing and collecting of vegetables and hunting being possible. The consequence was a high prevalence of malnutrition and famine.

The conditions in the camps concerning food production, processing and supply were, and still are, totally different, and can be summarised as follows:

- Cultivating rice is not possible, and inhabitants depend on external supply.
- Cultivating and collecting vegetables is restricted, and often not possible.
- Raising livestock is not possible.
- Fishing is restricted.
- Collecting spices is restricted.
- Food and supplementary food is delivered mainly by NGOs.

The consequences are a low prevalence of malnutrition, but a high-to-total dependency on food delivery (Cusano 2001; Spivey and Lewis 2016).

The camps are overseen and run by The Border Consortium (TBC), a union of international non-governmental organisations, supported by up to twenty governments/donors that provide food, shelter and non-food items to up to 150,000 refugees and displaced people residing in nine camps on the Thai–Burmese border. The items in the food basket provided a total calorific contribution of 2,230 Kcal per person/day, which had been well accepted by the refugees. However, this was reduced in 2013 to encourage the refugees to return to Myanmar. People with particular needs, such as malnourished children, pregnant and nursing mothers, the aged and the sick, are met by the nutritional feeding programmes run by health agencies. There is also nursery school feeding and regular monitoring of the nutritional status of the refugees in general (The Border Consortium 2013).

How Dependency Can Develop

First step: The first documented refugees from Myanmar arrived in Thailand in 1976, and were dispersed between several small so-called 'displaced persons camps' along the Salween river that forms the border. These camps each held between 300 and 2,000 refugees, who made their living by trading goods. At first, aid agencies provided essential drugs, vaccines, and basic health-care training and services, as well as certain basic commodities.

Second step: During the mid-1980s, the refugees were able to earn their own income, and retained control over their housing and most of their food supply. They were able to plant their paddy fields and vegetables across the border in Myanmar and to raise domestic livestock in the camps. Assistance was minimal, and was mainly organised and managed by the refugees themselves.

Third step: The large increase in the number of refugees entering Thailand after 1988 and again in 1994/95 resulted in a more systematic 'top-down' approach. Health care, shelter and nutrition were provided, with planning and implementation mainly through NGOs, who were requested by the Royal Thai Government to increase their services in order to avoid outbreaks of disease. These services included: implementation of a health surveillance system; provision of essential drugs; immunisation against communicable diseases; treatment of the most problematic diseases such as diarrhoea, malaria and tuberculosis; laboratory training and services; training of refugees in health-care services and management; and water supply and sanitation. The NGOs also needed to provide food supplies and shelter, as the refugees were no longer allowed to organise their own. The level of humanitarian assistance was not allowed to exceed the living standards of the Thai host communities, in order to avoid inequalities.

Fourth step: There was a consolidation of the camps in the late 1990s, result-ing in larger camp settlements, with up to 45,000 refugees in the largest camp. The number of camps was reduced from twenty-nine in 1994 to nine camps by 2007. Additional stringent movement restrictions, set by the host govern-ment, resulted in increased confinement in the camps with limited work and educational opportunities, which has led to almost complete dependence on aid over the last five years.

The central problem was the neglect of refugee participation. As Dr Benner of the NGO, Malteser International, concluded: 'The refugee participation shifted from self-reliance for shelter and food to the current situation in which the refugees have become fully dependent on the international community for their living in Thailand, tempered by partial self-management of their own health care, education services and food distribution' (Benner et al. 2008: 25).

Sociocultural Considerations

Having lived for many centuries in small villages of up to thirty or forty families, and having survived on small-scale animal husbandry, cultivation of several sorts of rice and different types of vegetables, as well as hunting and fishing and slash-and-burn agriculture, the Karen have now had to face an unfamiliar situation after their forced migration to the refugee camps and their lives have changed dramatically and more or less completely. The

Figure 9.2 Street scene in a long-term refugee camp, Mae Ra Ma Luang on the Thai–Burmese border, 2017. Photograph by the author.

new locations they have had to live in are similar to each other, being crowded places with each campsite hosting more than ten thousand inhabitants. The higher degree of anonymity forced people into more cooperation with their in-groups, like extended families or new neighbours. Organisation structures, which in the past had not been necessary to develop to this extent, evolved, like the Karen Women's Organization (KWO) and camp committees. Interaction with NGOs also produced new structures, replacing the ones lost in their villages of origin.

Family problems in the camps are now not so easily hidden as in the past, and domestic violence has become more a subject of public discussion. Negative coping with stress and the feeling of uselessness (because of having no work to do), helplessness (because of being dependent on the support of third parties and not being able to support oneself or one's family on one's own), and hopelessness (concerning one's future and that of one's relatives, the situation in the camps and the developments in Myanmar in general), probably increased the danger of domestic violence and the abuse of alcohol. Even when the NGOs have tried to establish some educational and health standards in the camps, it has been hindered as some of the camps are located in remote areas with no access to internet, television, telephone or other kinds of media and systems of communication.

Having been dependent on the help and support of NGOs and other organisations, and on the political will of the Thai and Myanmar governments for more than two decades, the refugees are probably finding it increasingly difficult to cope with a now changing situation caused by the pressure to be repatriated back to Myanmar or resettled elsewhere. Resettlement in general

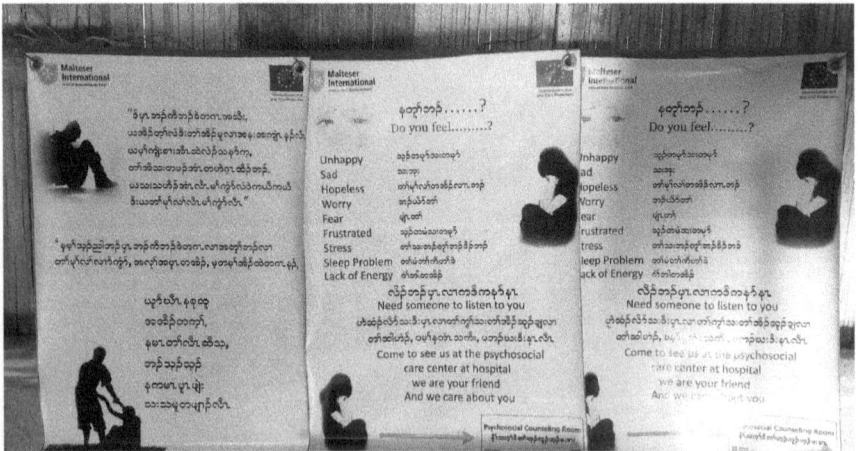

Figure 9.3 Mental health education poster displayed on the wall of a hospital at Mae Ra Ma Luang on the Thai–Burmese border, 2017. Photograph by the author.

can only take place for refugees who are registered, which is only possible for those refugees who left their country and arrived in the refugee camps before 2006.

Regarding the social and cultural background of the Karen refugees, one can ask why suicides and the clinical relevance of psychosocial problems could not have been observed more often over the years in the camps. Available data on the suicide rate for some years, as well as an estimation of the number of suicides during other years, make it probable that the suicide rate is one of the highest in the world. Moreover, amongst the Karen, as in many other Asian ethnic groups, psychosocial problems, stress and suicidal thoughts are taboo subjects, resulting in an underreporting of mental health and of problems associated with psychosocial conditions.

Fortunately, humans are very adaptable to changing living environments, especially when young. Now nearly all of them have been born in the camps and therefore do not know how life would have been in the land of their ancestors. Beside school education, the NGOs over the years have tried to help the refugees to retain their knowledge about a self-supported, autonomous and more or less self-sufficient way of life. An effort has even been made to introduce new production techniques of a (more) sustainable character, such as, for example, soap production. This initiative was set up initially in the camps, and the acquired knowledge was eventually distributed to surrounding Thai villages, helping to build up the capacities in these communities too.

The greatest obstacle for the refugees returning to their home country or being resettled is to face the fact that, from then on, they have to be self-reliant; they have to stand on their own feet. This special condition of 'learned helplessness' was the result of the official and unofficial Thai policy of not encouraging refugees to settle permanently in areas that are more or less similar to the natural environment from which they had come and which up to now had not been densely populated. These areas are also on Thai soil and therefore off-limits for people from Myanmar (Kaiser 2005).

So, I suggest that there are lessons here that could be learned by NGOs, as well as by governments.

Lessons from these Experiences

This refugee population has moved from relative independence in the early years to an almost total dependency on aid. The refugee 'participation' has been reduced to providing human resources for health and education services and food distribution. The population is more responsible for the administration of activities than for the design and planning of programmes. If this is to be avoided both in this situation and in other protracted refugee crises, the international community and host governments need to pay far greater attention to:

- involving refugees early on in the planning.
- designing programmes that provide work opportunities to ensure self-sufficiency and to reduce aid dependency.
- ensuring that aid supports the integration rather than the isolation of refugees, with an emphasis on building trust, synergy and good relationships between refugee and host communities.

Discussion

The question that has to be asked is this – is sustainability a luxury? Sustainability will continue to be a subject of discussion in the future, but the subjective psychological resistance and power (so-called resilience) of concerned populations and individuals will be the focus of research too. How people are able to cope with crisis is interesting from a psychological point of view, especially as unstable living conditions will probably become more common than stable ones in the next few decades; perhaps they will even become the norm. Despite early warning systems, the incidence of man-made and natural disasters will no doubt increase, and coping with them will become a daily reality for much of the world's population. These events will permanently penetrate our consciousness, not only because of the presence of the media but because of their global impact.

A push for the intensification of sustainability and resilience should not distract us from the fact that, for the majority of people in poor countries, their daily survival and that of their families is a higher priority than the longer perspective of sustainability. Sustainability has to do with responsibility towards the nature that nurtures us. However, in the case of the Karen described above, it is the human who comes first, and responsible thoughts about nature and its preservation come second. As long as the living conditions of individuals, ethnic groups and people are endangered, even if that is only in their subjective perspective, nature suffers too. That there are ways in which some of these things can be solved can be seen in the examples of training young people in the camps in sustainable animal husbandry and agricultural techniques. It remains to be seen, however, whether these young people are able to preserve their newly acquired knowledge after resettlement or repatriation back to Myanmar.

The FAO's definition of sustainable diets reminds us of the need for a more mindful approach to the care of the earth and its resources, but it will still be a long way to go for those people and ethnic groups whose living conditions are currently far from safe and sustainable. Fighting for daily survival turns food sustainability into a luxury. 'Our main need is security. We want the Burmese soldiers out of our area, taking the landmines they planted out too. When we return, we don't want the Burmese soldiers harassing us again' (The Border Consortium 2008b).

Conclusions

The widespread notion of the close proximity of nature in the livelihoods of indigenous peoples and their sensitive and sustainable handling of the resources at their disposal is perhaps a concept emanating from those living in urban habitats far away from so-called nature. As has been argued, sustainable management of food resources in some socio-economic settings is not a question of consciousness, but of priorities. Sustainability generally means avoiding the over-exploitation of animate and inanimate natural resources and dealing gently with the available resources. This often implies more elaborate processing and more restrained use of food. In emergencies, these considerations are either not employed or play a subordinate role in deciding which resources are used or exploited, and to what extent. The protection of natural resources and the best use of arable land, as well as the appropriate use of foodstuffs of all kinds, either require corresponding knowledge or the possibility to implement appropriate knowledge. Sustainability in the sense described above is thus subordinated to the acute survival of the group. Or, due to a lack of knowledge, it is not a central aspect of the employment of the affected persons in dealing with the food resources surrounding them.

In summary, therefore, a sustainable, healthy food supply and intake can be threatened by:

- lack of (or loss of former) knowledge of food processing procedures
- lack of possibility to implement existing knowledge (e.g. not enough time during flight for food preparation procedures)
- lack of resources (e.g. energy for proper cooking and destroying harmful ingredients).

Peter Kaiser
Swiss Red Cross Center for Victims of Torture and War, Switzerland

Notes

1. A refugee is a person who is outside their country of citizenship because they have well-founded grounds for fear of persecution because of their race, religion, nationality, membership of a particular social group or political opinion, and is unable to obtain sanctuary from their home country or, owing to such fear, is unwilling to avail themselves of the protection of that country (UNHCR Refugee Convention 1951).
2. In contrast to refugees, internally displaced people (IDPs) have not crossed a border to find safety. Unlike refugees, they are on the run at home. While they may have fled for similar reasons, IDPs stay within their own country and remain under the protection of its government, even if that government is the reason for their

displacement. As a result, these people are among the most vulnerable in the world (Internal Displacement Monitoring Centre 2015).

3. According to the causes of population movement, forced migration may be an armed conflict-induced displacement: this is either as a direct consequence of fighting and counter-insurgency operations, or because armed conflict has directly undermined human and food security, and is linked to severe human rights abuses (Internal Displacement Monitoring Centre 2015).

4. The official English name was changed by the country's government from the 'Union of Burma' to the 'Union of Myanmar' in 1989, and still later to the 'Republic of the Union of Myanmar', which since has been the subject of controversies.

5. This consortium was called The Thailand Burma Border Consortium in 2004 and following years, but changed its name to The Border Consortium in 2012.

References

Asian Human Rights Commission (AHRC) (1999) Report of an Internally Displaced Man. In AHRC *Voice of the Hungry Nation: People's Tribunal on Food Scarcity and Militarisation in Burma*, Hong Kong, p. 24.

Benner, M.T., Muangsookjarouen, A., Sondorp, E. and Townsend, J. (2008) Neglect of Refugee Participation, *Forced Migration Review*, 30: 25.

Burlingame, B. and Dernini, S. (eds) (2010) *Sustainable Diets and Biodiversity: Directions and Solutions for Policy, Research and Action*, FAO, Rome.

Cusano, C. (2001) Burma: Displaced Karens: Like Water on the Khu Leaf. In Vincent, M. and Refslunc Sorensen, B. (eds) *Caught between Borders: Response Strategies of the Internally Displaced*, Pluto Press, London, pp. 138–171.

Human Rights Watch (2012) Ad Hoc and Inadequate: Thailand's Treatment of Refugees and Asylum Seekers, *Human Rights Watch Report*, September 2012. Published at: http://www.hrw.org/sites/default/files/reports/thailand0912.pdf. Accessed on 19 December 2013.

Internal Displacement Monitoring Centre (IDMC) (2015) Global Overview 2015, People Internally Displaced by Conflict and Violence. Published at: http://www.internal-displacement.org/assets/library/Media/201505-Global-Overview-2015/20150506-global-overview-2015-en.pdf. Accessed on 16 November 2016.

Kaiser, P. (2005) Coping with Stress, Mental Health and Psychotherapy in Long-Term Refugees, *Viennese Ethnomedicine Newsletter*, 8(1): 8–15.

Marshall, H.I. (1922 [1997]) *The Karen People of Burma: A Study in Anthropology and Ethnology*, White Lotus Press, Bangkok.

Renard, R.D. (1979) Kariang: History of Karen–T'ai Relations from the Beginnings to 1923. Publications of the University of Hawaii, pp. 37f., 46f.

South, A. (2007) Towards a Typology of Forced Migration in Burma: Burma's Displaced People, *Forced Migration Review*, 30: 16. Published at: http://www.fmreview.org/sites/fmr/files/FMRdownloads/en/FMRpdfs/FMR30/16.pdf. Accessed on 16 November 2016.

South, A. (2008) *Ethnic Politics in Burma: States of Conflict*, Routledge, Abingdon.

Spivey, S.E. and Lewis, D.C. (2016) Harvesting from a Repotted Plant: A Qualitative Study of Karen Refugees' Resettlement and Foodways, *Journal of Refugee Studies*, 29(1): 60–81.

The Border Consortium (2013) Refugee and IDP Camp Populations: April 2013. Published at: https://www.theborderconsortium.org/media/11722/2013-04-apr-map-tbc-unhcr-1-.pdf. Accessed on 18 November 2016.

The Border Consortium (2014) Program Report January–June 2014. Published at: http://www.theborderconsortium.org/media/51534/2014-6-Mth-Rpt-Jan-Jun.pdf. Accessed on 18 November 2016.

The Thailand Burma Border Consortium (2004) *Between Worlds: Twenty Years on the Border*, The Border Consortium.

The Thailand Burma Border Consortium (2008a) Internal Displacement and International Law in Eastern Burma. Published at: www.tbbc.org/idps/report-2008-idp-english.pdf. Accessed on 9 December 2013.

The Thailand Burma Border Consortium (2008b) (14) Karen Man, CIDKP Focus Group, Palaw Township, interviewed in June 2008.

UNHCR (1951) The 1951 Refugee Convention. Published at: http://www.unhcr. org/1951-refugee-convention.html. Accessed on 16 November 2016.

UNHCR (2015a) Thailand. Published at: http://www.unhcr.org/pages/49e489646.html. Accessed on 16 November 2016.

UNHCR (2015b) Myanmar. Published at: http://www.unhcr.org/pages/49e4877d6. html. Accessed on 16 November 2016.

CHAPTER 10
IN PRAISE OF A FERMENTED BREAD: AN ETHIOPIAN RECIPE FOR FRUGAL SUSTAINABILITY

Valentina Peveri

Introduction: The Paradox of Frugal Tastes

Consider the following paradoxes of plenty and scarcity: customers in wealthy areas of the developed world, who have educated themselves about 'healthy' food, may increasingly decide to favour brown rice, quinoa, amaranth and millet over white flour because of the perceived dangers of wheat and the epidemic of (frequently self-diagnosed) gluten intolerance. The rise of low-carbohydrate 'palaeo diets' has instigated scientific and pseudo-scientific debates about the bright and dark sides of minimalist diets. Its advocates argue that variety has been grossly overrated, whereas detractors reaffirm the benefits of 'eating the rainbow' and advise people to incorporate more fruits and vegetables into their diets. So-called 'Foodies' all over the world, or at least those who can afford to seek and pay for new culinary experiences, seem to pander to the qualities and tastes of exotic grains. Nowadays some very wealthy individuals are rediscovering the charm of poor people's food. Interestingly enough, one of the contemporary phrases that gourmet circles use to market these newly sourced 'miracle foods' (such as einkorn, kamut, fonio, quinoa, millet, sorghum and teff) is 'ancient grains'. In this way, some foods are going out of fashion, others are on the rise. West Africans are eating more like Asians; Asians are eating more like Americans; and the richest Americans have started fancying a few selected (and palatable) African staples (*The Economist* 2017).

These gastronomic riddles show how easily, in a cosmopolitan world, many (but not all) people can switch to different diets. They make choices based on reflections that have to do with agricultural technology, work, health and social aspirations, with being or feeling 'green', with what they may define as 'sustainable' consumption and/or lifestyle, or simply to add a pinch of culinary wackiness to an already chameleonic personality. It is not the scope of this chapter to delve into the intricacies of these food fads and labels, nor to determine whether there is value in variety over monotony, or the other way around. Nor shall I solve the dilemma of which foods are the more virtuous or sustainable. My goal is instead to read between the lines of these circular, sometimes paradoxical food choices, and highlight three concepts that strike a delicate chord with an anthropologist who grew up professionally in a rural, subsistence-oriented community in south-western Ethiopia, where low-intensity practices dominate the field, and the devotion to local resources prevails in the kitchen and at the table. The three concepts to be discussed are subsistence, sustainability and frugality, after which original fieldwork material will be given on the production, preparation and consumption of the ensete bread staple of the Hadiyya people in Ethiopia. This chapter will then discuss the benefits of, even local preference for, frugal foodways, and seek to throw light on some misconceptions about frugality and how it achieves sustainability.

Subsistence

In the circular tale of changing food preferences, the existence, let alone the value, of self-sufficient and subsistence-oriented communities frequently goes unrecorded. Descriptive and analytical priority is given to those who move, geographically or metaphorically, within the global, interconnected and hybridised food landscape; while an incredibly loud silence enshrouds those who, by choice or under duress, still hold to their places of origin, are not forced to change through migration or cultural disruption, and basically struggle to retain their familiar tastes and flavourings. Fonio or teff, so to speak, are more likely to become fashionable in a well-fed population when manipulated by celebrity chefs and beautified in gastro pubs rather than in the daily and unspectacular preparations of (female) African cooks. 'Haute cuisine' as expressed in restaurants or ethnic eateries is but one aspect of culinary history. Yet, a reference book on the populations that nurture a strong attachment to traditional staple foods has still to be written, because the fact that 'human beings are remarkably conservative in their food habits and are typically reluctant to try new foods and to abandon old, familiar ones' (Rozin and Rozin 2007: 35) would hardly capture the imagination of readers or consumers. In the 'Global North', it is consumers who drive demand and changes in diet, consumers who have acquired a taste for novelty as urban workers or gastrotourists, while government policies are increasingly shaped within neoliberal rationalities and corporate interests. In the same vein, the bulk of development literature envisions a transition from mostly subsistence farming to high-value commercial farming; such rhetoric relies upon the axiom that subsistence is tantamount to poverty and, as a consequence, to lack of success and empowerment of poor farmers, who would, therefore, require an integrated suite of services to grow cash crops and break out of subsistence production.[1] The conventional wisdom is that small family farms are backward, conservative and unproductive. On the contrary, I raise the following question: Is there any counter-evidence to this of the potentially transformative role of subsistence agriculture, one that contains seeds of sustainable alternatives from field to plate?

Sustainability

In the present political and cultural climate, the language of sustainability is at the same time ubiquitous and fuzzy. It spills over into the media, environmental policy, academic agendas, and popular imagination, while it is used to justify a plethora of phenomena and practices, such as vegetarianism and animal rights, movements towards slow and mindful food choice and consumption, human rights and social justice (Evans 2011). Concerns about personal health blend into reflections about the eco-cultural health of the environment (Rapport and Maffi 2011). In such societies, the concept of frugality or 'voluntary simplicity', through choice or necessity, in order to reduce environmental impacts but also to save money and therefore bordering on thriftiness, is sometimes evoked. Yet, that frugality typically refers to consumption patterns in industrialised countries, and to individual choices of, say, buying eco/environmentally friendly or fairly traded alternatives, and similar attempts to moderate an otherwise unrestrained consumer lifestyle. Whereas this frugality, supported by concepts perhaps of global sustainability, becomes an individualised remedy to overcome feelings of guilt or a minimal act to resist the frantic need to consume, it has obviously been tailored on the ethical needs and sensibilities of Westernised consumers. But is it the only version available for framing sustainability in the twenty-first century? Or do we have other angles to look at frugality in a more politically engaged and inclusive way?

Frugality

As theorised above in relation to those affording options, frugality takes on the meaning of a morally laden, deliberate choice. However, when referring to peasant societies or 'poor' people's food habits, frugality is frequently related to concepts of scarcity and hunger; even more so if the reality under observation happens to be in sub-Saharan Africa. This biased perception remains particularly ingrained in Western portrayals of the continent, and is disproportionately prolific in media images of global hunger as well as media-based humanitarian campaigns. On the contrary, ethnographic research has demonstrated that African people, particularly in regions that face repeated shocks, have learned both to plan for contingencies and to respond actively to them – by diversifying, rotating and planting drought-resistant crops, substituting bulkier foods for others, fattening up or periodically fasting, cutting down meal size and number, concealing consumption from the public gaze, and/or maintaining friendships with distant groups practising other modes of livelihood (Shipton 1990: 363–364; Messer and Shipton 2002: 231). The ingenuity of subsistent farmers to solve practical issues through a fine-tuned deployment of coping strategies – from self-rationing to conserving resources through lean seasons – flies mostly under the radar of wider public attention and appreciation.

I argue that we have thus turned full circle within the initial conundrum of frugal tastes: Why are we prone, as researchers and food citizens, to use two different yardsticks in pondering frugality, when we frame it as meagre scantiness 'abroad' and non-wasteful sufficiency 'at home'? Whenever a commentator suggests that a return to frugality represents a unique opportunity for the pursuit of sustainable consumption, and points to societies where economic activity is consistent with a holistic ethic of care (towards humans and non-humans alike), those examples of sustainability are typically referred to as 'moral economies of the past' (Evans 2011: 556). Yet these frugal alternatives do not belong to a remote past and are in fact fully contemporary. I shall now briefly overview how frugality is conceived and practised outside the comfortable territory of the Western paradigm with reference to my own fieldwork.

The Ethiopian Bread Cake: From Field to Table

I use evidence from fieldwork in Hadiyya in south-western Ethiopia to illustrate through a palpable example of what meals look and taste like in a self-sufficient farm. The landscape, where I sharpened my ethnographic focus, is dominated by ensete (*Ensete ventricosum*) – a plant that has, over centuries, come to be called the 'tree of the poor' or 'tree against hunger' because of its resilience and long shelf life as a fermented food during famines. The pattern of simplicity among these African farmers will be briefly analysed by detailing the preparation techniques of a fermented food similar to the Elven bread (Lembas), which Frodo and Sam had to parcel out so frugally: 'Eat little at a time, and only at need. For these things are given to serve you when all else fails. The cakes will keep sweet for many, many days if they are unbroken and left in their leaf-wrappings, as we have brought them. One will keep a traveller on his feet for a day of long labour, even if he be one of the tall Men of Minas Tirith' (Tolkien 1984: 390).

My doctoral and postdoctoral fieldwork[2] has revolved around ensete. It is only in Ethiopia that ensete has been domesticated. Ensete is not widely known as a food plant outside of its place of origin, and the culture of ensete remains obscure internationally relative to the size of the populations that subsist on it. Ethiopians in most areas of the south-western highlands have cultivated this monocarpic, perennial plant, a cousin of the banana tree, for thousands of years, and have converted its stalk, roots and leaves into food, medicine, decoration, and more.[3] Both ensete and bananas are tree-like, giant, herbaceous plants, but ensete flowers only once during its final stage (at around the age of ten years) and bears fruits that humans do not consider edible, even in times of scarcity. The part of the plant that is used for human consumption is not the fruit, but the enlarged pseudostem and underground corm, with a circumference of up to 2.5 metres at maturity, which swells over time with carbohydrates. Ensete has persisted through droughts and floods by storing water in its bulbous stalk; it shields the soil from erosion with its long-lasting roots and renders the land rich in nitrogen, water and humidity; and it thrives especially when fertilised by cattle manure rather than by chemical inputs.

Ensete is prepared in a variety of ways. I had the privilege of being nurtured by my host Hadiyya families from the very beginning with the main products of the ensete plant – *waasa*,[4] *bu'o*[5] and *ha'micho*.[6] Whereas *ha'micho* is steam-boiled and immediately cooked, the starchy fermented paste of *waasa* and *bu'o*, after an elaborate process to extract pulp from leaf-sheaths, is steam-baked – *waasa* in the form of unleavened bread, *bu'o* as porridge or gruel to which milk and butter can be added. People occasionally consume these products immediately after harvest, but fermentation of ensete pulp for months or years is the norm. The pulverised corm and squeezed leaf-sheath pulp are mixed and then wrapped with several layers of ensete leaves as pre-fermentation procedures. The resulting substance is stored safely underground to ferment, with periodic remixing and kneading until the material is considered properly fermented. Behind the house, numerous underground pits lined with leaf sheaths hold ensete pulp while it ferments. The ability to store processed ensete with little storage loss for long periods of time (from a minimum of two weeks up to ten years) provides households with a mechanism to modulate consumption during food shortages.

One of the most remarkable characteristics of this political ecology lies in the fact that the landscape and foodscape of ensete are dominated by the women's great knowledge of, and work with, the plant. Ensete is categorised into various stages of growth based on the age, size and perceived gender of each plant. There are over two hundred vernacular names for the ensete plant in Ethiopia; differences in name are related to differences in the utilisation of each variety. Women grow ensete in controlled and well-regulated spaces (home gardens) that teem with, and generate, high levels of agrobiodiversity. They bring in and mix materials from various sources, such as from neighbours, friends, relatives and/or from markets; and through the same, mainly informal networks they save and exchange ensete cuttings. For ensete processing, several women from the community are recruited by a household and gather to perform this highly engaging task. Work is carried out collectively and through a generational division of labour, in small groups where little and adolescent girls learn through the processes of observing, asking questions to, and mimicking the gestures of, elderly women. When the ensete plant is mature and ready to be dug out, all the external leaves and dried parts are cut and the remaining inner structure of the plant – similar at that point to a polished, shiny white column – is thoroughly cleaned; then the roots are cut off from the base. With some light shaking back and forth, and a final, vigorous push, the plant falls (Figure 10.1). This is the point beyond which no exception is made to the presence in the garden of men. Under no circumstances would they be allowed to touch the tools or substance of ensete, which are handled exclusively by women. The harvesting activities, together with the exchange and barter of raw or processed ensete materials, contribute across seasons to reinvigorate community feelings. Through donation and reciprocal obligation, women control the distribution of garden foods outside the family, and create networks of alliance between households.

More importantly, ensete growers have traditionally engaged in wide-ranging, and fine-tuned, combinations of livelihood activities to lessen the risk of crop failure and food insecurity, including cereals, legumes, root crops, fruit trees, livestock, stimulants, timber, off-farm work, trade and the foraging for wild foods and seasonings. They have known for generations to plant ensete alongside a diverse range of crops that provide different nutrients, help to offset the boredom of eating the same food every day, and can occasionally be sold at the market. They carefully stagger planting times so that plants mature and are available throughout different seasons, sometimes coordinating with kin and neighbours in order to maximise access through sharing. In this way, the garden acts as a living pantry, providing fresh produce on a daily basis.

Being involved at every stage from production to preparation to consumption, Hadiyya women are guided in their performances by a wide range of criteria, such as agro-ecological adaptation, length of the digestion time, swelling capacity during cooking, taste, culinary uses, and post-harvest characteristics. In turning ensete into palatable products, the quality of the ingredients is a prerequisite for the perfect dish; but the amount of labour spent in the garden to grow, tend, process and transform the plant is just as important. For a long time, I thought I had acquired a full and refined taste for ensete bread, but years later I found out that most of the locally perceived characteristics (such as texture and visual appeal) had in fact eluded my culinary sensitivity. In 2014, during a long and intense period of fieldwork to collect as much data as possible on local ingredients and recipes, women decided to share with me three different bread recipes. As documented in other African peasant societies, 'preparation of the simple meal is lengthy and laborious' (Spittler 1999: 33). The post-fermentation steps require several hours of preparation of the material before it can be cooked; every step is essential as it will greatly affect the taste of the final dish. There is a common pattern to follow: extract small portions of fermented pulp, wrap them separately, and leave them to dehydrate for a few days; remove scraps and impurities

Figure 10.1 The ensete plant being harvested. Photograph by the author.

by repeated kneading and chopping with a knife, grate the resulting paste with a stone mortar and shape it into a round cake with a diameter of 40–50 cm and a thickness of 2–3 cm (Figure 10.2). The paste, wrapped in fresh ensete leaves, will finally be baked on a large, flat griddle over a medium fire for about an hour. Variations in the quality of bread occur depending on how mature (dried) or young (fresh) the pulp is, whether fibres are kept in, chopped into small pieces or completely removed, whether or not it requires to be soaked in water, and finally whether the resulting taste will be strong or weak, the texture soft or resistant, and the crumb's colour verging on either brownish or glossy white. The basic flavour may be varied with the addition of seasoning ingredients. The bread is finally paired with different sauces, most typically hot spicy vegetable ones. If necessary, it can be eaten alone because, insiders argue, it gives warmth, energy, a sense of repletion and culinary pleasure. The bread will remain fresh for over a week and constitute a substantial meal for all family members. The Hadiyya women farmers have repeatedly expressed to me the benefits of ensete in their lives, and explained that fifty ensete roots, which yield up to 40 kilograms of food, could feed a family of five or six for several months.

In my long and intensive exposure to this ensete culture I have slowly realised that ensete is not a unique miracle plant, and that, moreover, it fosters an alimentary experience laden with geopolitical power. Low-intensity cultivation practices characteristic of ensete mirror those for other 'orphan crops' (such as millet, yam, cassava, cowpea and sorghum), which might be defined as a diverse set of minor crops that are simple and cost effective for poorer countries. These crops are not traded around the world and receive little or no attention from research networks, but nonetheless play an important role in regional food security. Yet, until recently, government and development agencies have consistently failed to include in their nutrition and agricultural policies any appreciation of ensete food culture as a prized regional variation, or as a local flavour to be preserved and disseminated. At present, ensete farm families remain invisible from the public gaze as innovators in food production/processing. This invisibility – internationally and within national boundaries – clashes with the evidence that today 18 to 20 million people, 20 percent of the total population of Ethiopia, depend on this plant either as a staple food or as a famine crop. These populations have never starved, even during the tragic droughts of the 1970s and 1980s (Negash and Niehof 2004).

In a state where policy makers, scientists and plant breeders have periodically turned to high-yielding crops (preferably grains such as wheat and maize) to address the problem of food insecurity, instead of committing themselves to exploring the potential of traditional diets and of 'poor' man's foods; and where the official and backstage policies envision a move away from subsistence agriculture towards specialisation and market-based agricultural production (Peveri 2016), there is no room for Lembas bread to sustain Hobbits on any kind of epic quest. Extremely marginal value is assigned to the rural periphery. In examining some of the prospects discussed for the future of sustainability, we should ask: What alternatives might be excluded as a consequence? Who gains and who loses?

Figure 10.2 Preparing the ensete bread. Photograph by the author.

Discussion: No Frills Cuisines

The lifestyle of frugality among the Hadiyya described above is by no means exceptional. In the pioneering work of Audrey Richards, we find a pattern of simplicity in the way the indigenous people of Northern Rhodesia (now Zambia) expressed their views on food: 'In general, only one relish is eaten at a meal. The Bemba do not like to mix their foods and despise the European habit of eating a meal composed of two or three kinds of dishes ... One said, "It is like a bird first to pick at this and then at that, or like a child who nibbles here and there through the day"' (Richards 1995: 49). The Bemba thus compared idealised simple food as the 'perfect meal'. This pattern of simplicity has been documented in several other instances (Spittler 1999). However, the concept of simplicity, like frugality, raises contrasting views.

The simplicity of, and little variation in, African foodways have been frequently translated by external observers into the disparaging concepts of insipidity and monotony. According to this banalising perspective, African meals have the qualities either of a poor man's food or of a primitive, underdeveloped cuisine. The simplicity of 'the others' is conflated with the 'taste for necessity' (Bourdieu 1984: 168). In contrast, from a local point of view, simple diets are usually considered excellent and are praised for the delicious potential of simple ingredients transformed by a skilled cook (in most African contexts, a woman). To local people, the value of simplicity is often integral with, and triggers, the ability to manage landscapes of scarce resources, and turn such areas into places that produce foods that not only fill the stomach but also respond to the interrelated desires of healthiness, digestibility and good taste. This is not only true for African cuisines, but also sits well with what certain literature has sketchily defined as 'cuisines of poverty'. Even within Europe, a research project conducted among villagers in western Andalusia brings to light the high degree of dietary conformity in the lower classes who voluntarily opt for frugal day-to-day meals as a reminder of past times of food scarcity, and as a social marker to define who they are: essential, free from impulses for over consumption, traditional, territorial and non-elitist (González Turmo 1997).

The special issue of *Food, Culture and Society* (2017) entitled 'Less Palatable, Still Valuable: Taste, Agrobiodiversity, and Culinary Heritage' collects together case studies of people from across the world who commit themselves to cultivating and eating food plants that are perceived to be socioculturally and affectively important, even if the resulting taste on the plate may be described as bland, monotonous or unappealing. Edible but less palatable plants, it is suggested, are nonetheless greatly valued because 'any mention of "good" food – even when intended to be "merely" about flavour or health – comes with social baggage' (de St. Maurice and Miller 2017: 195). At the intersection of taste and value, as both driving factors in determining what should be cultivated, eaten, saved and maintained, these underrepresented crops shed light on how certain local foods may be considered good long before they reach the cooking pot or the baking oven: they may be good in the field for they contribute to agrobiodiversity; good in the belly as they make people feel healthy, strong, warm and food secure; and within the whole social fabric for their symbolic relevance and ritual applications. This is the case, for example, with processing and eating cassava and yams among the Canela of north-eastern Brazil (Miller 2017), or with praising the monotony of a sago-based diet among the Abelam of Papua New Guinea (Scaglion 2017). I found the same strong devotion to a single plant, and to its basic products, expressed by Hadiyya women as follows: 'People here in the village do not eat *waasa* out of necessity, but because they prefer it to other foods, even if the latter are available. If farmers eat *waasa* it will be digested after a long time, because it gives warmth and energy, and so they will not be hungry'. The consumption of ensete is explained not only in terms of a sense of repletion, but also in terms of healthy eating: *waas' nakkoyo*, they reiterate – 'you cannot get sick with ensete bread'.

Individuals and communities spend a considerable amount of effort and time to growing, harvesting, processing and cooking foods that would appear simple, nutritionally poor or ugly to an outsider. More broadly, studies of consumption habits and culinary lifestyles in the Global North tend to view tastes in isolation, rather than in connection to values and performances that may affect environmental sustainability at large. But taste is not everything, and 'when a crop is widely understood to have nutritive benefits or to be a reliable resource during lean times, people may grow it even when more easily obtained, more palatable alternatives abound' (de St. Maurice and Miller 2017: 197). Frugal cuisines are typically born out of a holistic understanding of human–environmental relationships. It is precisely in subsistence-oriented societies that it vividly emerges how strongly those relationships shape what comprises a 'familiar taste'. In self-reliant systems, food is not purchased but typically produced on a daily basis. People mainly subsist on locally cultivated staples, by carefully planning year-round agricultural labour. Farmers are physically and intimately involved in the precarious processes of cultivation and harvest. In those societies, the highly valued foods in terms of consistency, reliability and symbolic meaning are most likely the ones that are known and trusted locally (Richards 1995: 47).

To grasp the culinary and political-ecological richness of subsistence/ peasant cuisines we should keep in mind a pattern of culinary simplicity that has proven to be recurrent over time and place, where a core staple (in the form of porridge or bread) constitutes the foundation of a meal, is consumed in tandem with a variety of fringe foods (sauces, soups, stews), and peppered by the use of different condiments (Messer 2009: 55). The staple provides the dominant theme, while the side dishes bring into the composition regional, local, seasonal, or personal variation. The template may appear simple, but the realisation is complex and demanding, especially if we account for the whole trajectory from field to table. Moreover, 'what seems a monotonous repetition of flavourings may well be an illusion of the outsider ... A closer look at actual culinary practice reveals a rich and subtle variation of flavouring from dish to dish and from meal to meal' (Rozin and Rozin 2007: 40). In Hadiyya, as on other family farms, that cook has to feed several people at a time by using the limited seasonal ingredients on hand; and what we would simplistically frame as necessity or satisfaction of hunger has already become a brilliant choice that she plays with ingenuity and vision at every meal preparation. She may well be the same person who has grown and processed the raw items into food, and what is finally prepared and eaten would therefore demonstrate her personal engagement in production, the environmental diversity of her plot or garden, and the social climate surrounding her working and affective relationships.

From different varieties cultivated in the field to subtle variations in preparation, the ensete bread summarises some of the themes encountered through the discussions of frugality and sustainability above. First, contrary to any rash assumption of an ensete-based diet being gastronomically stagnant, what I have registered through my ethnographic lens is a cuisine far more colourful and vivid than most outsiders know or acknowledge. The misrepresentation of peasant cuisines seems, therefore, a matter of sharpening our vision to focus on the great subtlety and surprising variety simmering under the apparent simplicity. Secondly, what renders this simplicity dense and complex is that Hadiyya women, acting as experts both in the field and in the kitchen, within the contours of self-reliance and self-sufficiency, produce a bread that tastes not only 'good' or 'bad', but contributes to agrobiodiversity, a sense of belonging, and symbiotic interactions with the environment. Thirdly, the common narrative about necessity, loss and culinary absence in sub-Saharan Africa seems to foster a hindrance to our taking up the challenge of, and drawing inspiration from, radical forms of sustainability, wherein the frugality of ingredients and of culinary appetites unveils a sophisticated self-rationing system. Monotony does not equate with poverty, nor simple needs with deficiency, but rather with a natural alternation of plenty and shortage that individuals have learned to gamble on and adapt to. In sum, 'we are dealing here with a system of simple needs, in which people are content: with few possessions and a simple pattern of consumption' (Spittler 1999: 40).

Conclusion: Simple Meals for Sustainability?

In February 2008, as a result of several years of short or delayed rains, drought occurred while I was conducting my fieldwork. We ate very little, only a few mouthfuls each time, withdrawing from the pits where ensete was stored with respect and frugality. Neither I nor other people in the area starved to death.

We never felt that we were lacking food. Hadiyya farmers are used to stoically enduring periods of scarcity. As for my part, I had to learn very quickly, and painfully, how to slide into frugality mode. This episode was followed by other field stays, marked by fluctuations in scarcity and plenty, and created a fertile ground to start reflecting upon the capacity of common people to create value (and good meals) from limited resources.

This chapter builds upon that original reflection, and aims at opening up new lines of inquiry into the potential, and positive societal effects, of intertwining the concepts of frugality and sustainability within the public debate over food justice. What makes this connection even more relevant is the momentum that the concept of 'frugal innovation' is gaining in the engineering and hi-tech sectors (*The Economist* 2010). The phrase refers to innovations based on minimal use of resources for realising efficient and robust products that are sustainable for the environment and communities, based on simple design, and stripped of non-essential features (see, for example, the case of the bamboo bicycle).[7] Most of these innovations are born out of the ingenuity or creative improvisation of common people, and capitalise on their ability to quickly adapt to uncertain circumstances in an intelligent way (Pansera 2013; Khan 2016). This angle, if adopted in food practices, triggers the idea that good lives can be led without high levels of consumption.

Not all the intricacies of food sustainability will be resolved through frugal choices. Frugality and sustainability are an ongoing process, one in which narratives of sustainable choices accomplished by individuals in consumer cultures remain strictly interconnected with the rights of self-sufficient households to support themselves, and to shape their own blend of production; both navigate within wider structural limitations and political agendas. The critical role and responsibility of the state cannot be underestimated; for example, in revitalising green and resilient production systems, while at the same time expanding the rights, participation, self-confidence and well-being of farmers in subsistence agriculture. However, as the Ethiopian case suggests, a frugal mindset may prove highly adaptive for enduring constraints and learning how to do more with less. The case of the Hadiyya people, like many other small-scale farmers in developing countries, confronts us with the question of whether variety in food is the norm and it is the simple and unchanging that is a phenomenon requiring explanation; or, on the contrary, is it our need for alternating tastes that calls for investigation? More radically, the possibility might be suggested that producing differently, instead of consuming less, should be carefully considered and politically implemented in reasserting the value of sustainability along with food justice and social equity.

Valentina Peveri
The American University of Rome (AUR), Italy

Notes

1. By way of example, see *The Roadmap for Promoting Women's Economic Empowerment*. Available from: http://www.womeneconroadmap.org/roadmap [accessed 21 March 2018].
2. The first time I set foot on an ensete garden was in November 2004. Since then I have been conducting ethnographic research in Leemo District, Hadiyya zone, part of the Southern Nations Nationalities and Peoples' Region (SNNPR), paying annual visits of several weeks or months each year up until the spring of 2015.
3. Every part of ensete is used: to cover roofs and walls; to make containers, ropes, mats, bags and sieves; its strong fibers are an item of barter; dried stems fuel the fireplace; fresh leaves are used as serving dishes, as protective covering inside the pits where ensete is fermented, or to pave the ground where it is processed; to wrap ensete pulp, honey, tobacco, butter, crops, and also, in the past, newborn babies; to pack and carry goods to local markets; as fodder for animals; and to treat fractures and problems related to childbirth or abortion.
4. Fermented starchy mixture of pulverised corm, decorticated pseudostems, and squeezed leaf-sheaths pulp.
5 A small amount of water-insoluble starchy product, in the form of white powder, that may be separated from *waasa* during processing by squeezing and decanting the liquid.
6. The fleshy inner portion of the ensete corm.
7. The Bamboo Bike Project is run by scientists and engineers at The Earth Institute, Columbia University. The project aims to examine the feasibility of implementing bicycles made of bamboo as a sustainable form of transport in Africa. Available from: http://www.bamboobike.org/Home.html (accessed on 21 March 2018).

References

Bourdieu, P. (1984) *Distinction: A Social Critique of the Judgment of Taste*, Harvard University Press, Cambridge, MA.

Evans, D. (2011) Thrifty, Green or Frugal: Reflections on Sustainable Consumption in a Changing Economic Climate, *Geoforum*, 42: 550–557.

González Turmo, I. (1997) The Pathways of Taste: The West Andalucian Case. In Macbeth, H. (ed) *Food Preferences and Taste: Continuity and Change*, Berghahn Books, Oxford, pp. 115–126.

Khan, R. (2016) How Frugal Innovation Promotes Social Sustainability, *Sustainability*, 8(10): 1034, doi:10.3390/su8101034.

Messer, E. (2009) Food Definitions and Boundaries: Eating Constraints and Human Identities. In MacClancy, J., Jeya, H. and Macbeth, H. (eds) *Consuming the Inedible: Neglected Dimensions of Food Choice*, Berghahn Books, Oxford, pp. 53–65.

Messer, E., and Shipton, P. (2002). Hunger in Africa: Untangling its Human Roots. In MacClancy, J. (ed.) *Exotic No More: Anthropology of the Front Lines*, University of Chicago Press, Chicago, pp. 227–250.

Miller, T.L. (2017) Valuing the Bad and the Ugly: Tasting Agrobiodiversity among the Indigenous Canela, *Food, Culture & Society*, 20(2): 325–346.

Negash, A., and Niehof, A. (2004). The Significance of Enset Culture and Biodiversity for Rural Household Food and Livelihood Security in Southwestern Ethiopia, *Agriculture and Human Values*, 21(1): 61–71.

Pansera, M. (2013) Frugality, Grassroots and Inclusiveness: New Challenges for Mainstream Innovation Theories, *African Journal of Science, Technology, Innovation and Development*, 5(6): 469–478.

Peveri, V. (2016) Ghosts of Hunger: An Anthropological View of Agricultural Intensification in Southwestern Ethiopia. *PSAE Research Series*, 13, African Studies Center, Boston, MA.

Rapport, D.J., and Maffi, L. (2011) Eco-cultural Health, Global Health, and Sustainability, *Ecological Research*, 26: 1039–1049.

Richards, A.I. (1939 [1995]) *Land, Labour and Diet in Northern Rhodesia: An Economic Study of the Bemba Tribe*, LIT Verlag, Münster-Hamburg.

Rozin, E., and Rozin, P. (2007) Culinary Themes and Variations. In Korsemeyer, C. (ed.) *The Taste Culture Reader: Experiencing Food and Drink*, Berg, Oxford and New York, pp. 34–41.

Scaglion, R. (2017) Sago: A Disparaged but Essential Food of the Abelam of Papua New Guinea, *Food, Culture & Society*, 20(2): 201–215.

Shipton, P. (1990) African Famines and Food Security: Anthropological Perspectives, *Annual Review of Anthropology*, 19: 353–394.

Spittler, G. (1999) In Praise of the Simple Meal: African and European Food Culture Compared. In Lentz, C. (ed.) *Changing Food Habits: Case Studies from Africa, South America and Europe*, Harwood Academic Publishers, Amsterdam, pp. 27–42.

St. Maurice, G. de, and Miller, T.L. (2017) Less Palatable, Still Valuable: Taste, Crop Agrobiodiversity, and Culinary Heritage, *Food, Culture & Society*, 20(2): 193–200.

The Economist (2010) First Break the Rules: The Charms of Frugal Innovation. Published at: https://www.economist.com/node/15879359. Accessed on 21 March 2018.

The Economist (2017) Of Rice and Men: A Circular Tale of Changing Food Preferences. Published at: https://goo.gl/DGLgDA. Accessed on 21 March 2018.

Tolkien, J.R.R. (1984) *The Lord of the Rings*, Guild Publishing, Great Britain.

CHAPTER 11
THE INDIAN 'MEAT DILEMMA': MALNUTRITION, SOCIAL HIERARCHY AND ECOLOGICAL SUSTAINABILITY

Michaël Bruckert

When it comes to addressing food issues through the lens of 'sustainability', one specific foodstuff is usually pointed at: meat. Indeed, it has been argued for decades that grain cultivation is more efficient in terms of protein and energy than animal protein production. In particular, animals appear to be inefficient users of water and nitrogen (Smil 2002: 620). Furthermore, the livestock sector is regularly blamed for releasing a huge amount of waste (nitrogen again, but also greenhouse gases). In a global context of dwindling resources, more and more voices are raised against the economic and ecological unsustainability of the meat system, calling for a lower meat production and consumption worldwide (D'Silva and Webster 2010 1; Emel and Neo 2015: 2).

Nevertheless, it is obvious that global foodways cannot easily be altered by discourses based on scientific rationales, mostly originating in well-fed countries. Besides being entangled in economic and ecological factors, the choice for a specific foodstuff is also determined by physiological capacities, imbued by cultural, social and political dimensions and subject to embodied feelings of affection, sensuality, pleasure and rejection. In broad terms, food is more than just a set of nutrients. Meat, in particular, does not come down to a basic animal flesh to be produced and eaten by unspecified actors in generic conditions. Across the ages and the places, meat has always been 'polyvocalic and polysemic' (Leroy and Praet 2015: 208), endowed with myriad meanings, being at the same time esteemed and abhorred, revered and rejected (Fiddes 1992: 226). The many ways in which animals are tended, killed, and consumed are at the same time physically sensed and mentally apprehended, taking on material, affective and reflexive dimensions (Evans and Miele 2012: 308).

To explore the intersection between meat and sustainability, this chapter, based on extensive fieldwork carried out between 2012 and 2014, addresses the case of contemporary India. This country is often depicted either as being traditionally vegetarian or as experiencing a 'diet transition' (Pingali and Khwaja 2004: 2) – an expression basically pointing to the fact that, due to economic growth, global capitalism, urbanisation and industrialisation, meat consumption would soar in emerging countries. Often associated with the latter idea is the notion of a 'livestock revolution' (Delgado et al. 2001), whereby the change in food consumption patterns would go along with a rapid growth in stocks of farmed mammals and birds.

A few questions can be asked in order to unravel these two theories. Addressing the sustainability of meat circuits in India entails straddling the realms of moral economy and political ecology. In this chapter, I will first investigate the alleged rise in meat consumption in this country, thereby questioning the complex and contested values and meanings bestowed to this specific foodstuff. Secondly, going back up the supply chain, I shall explore the concerns about sustainability raised by meat production. Thirdly, I shall identify tensions or incompatibilities between the cultures of consumption and the structures of production.

The Cultures and Politics of Meat Eating in India

A quick look into statistics immediately reveals that, even if only 30 percent of the population is strictly vegetarian (Robbins 1999: 413), meat consumption is still very low in India. According to the figures provided by the National Sample Survey Organization in 2011, under the Ministry of Statistics of the Government of India, Indians eat an average of 3 to 4 kg of meat per person yearly, being 2 kg of chicken, 1 kg of mutton and less than 0.5 kg of beef,[1] whereas the world meat consumption average is around 43 kg.[2] The statistics show that the consumption of fish (3 kg) and of vegetable protein consumption (around 20 kg) is low as well.

Admittedly, discourses about nutrition can be understood as being both colonial (Arnold 1994) and highly normative (Hayes-Conroy and Hayes-Conroy 2016). But it is hardly deniable that these remarkably low intakes could well be a reason for the high prevalence of malnutrition and anaemia in India (Sébastia 2013), to the point that a 'protein deficiency' can be discerned.

Meat consumption varies across India: it is higher in the north-eastern and southern states, and extremely low in the north-western states of the country. This uneven geography does not pertain to economic factors and development indexes (wealth, urbanisation, literacy rate, etc.); rather, it is correlated with cultural factors, such as the dominant religion, the ethnolinguistic groups and castes that are present locally. Primarily, meat consumption is highest in areas where the percentages of Muslims, Christians, tribal populations or 'low-caste' members are also highest.

Beyond these variations, what are the factors that account for these low levels of meat consumption? Interviews carried out among the working classes and the middle classes in the meridional state of Tamil Nadu reveal that meat consumption is still highly regulated: the status of meat is defined by lingering conceptions about culture, religion, caste and health. These regulations, which would take too long to describe exhaustively here, are multifold. Nevertheless, they need to be briefly outlined.

First, there is an economic perspective, whereby consumers keep away from non-vegetarian food because meat, and especially mutton, is expensive. In addition to its cost, this product is also culturally still considered as a side dish: for many Indians, a proper meal must always include a huge amount of cereals, namely rice, wheat or, more rarely, millet.

Secondly, alongside this 'culinary' perspective, several religious and ritual restrictions also deter people from eating meat. With vegetarianism being a marker of purity and of superior status in the caste system, many Indians (especially Hindu Brahmins) still radically eschew meat. Furthermore, many Hindus from middle and high castes refrain from eating meat on 'auspicious days' (sometimes as often as two or three days per week for most of them, but also on new moon and full moon days), when visiting a major temple or when preparing for a pilgrimage. A temporary state of what is seen as a 'ritual defilement', for instance following the death of a relative or during menstruation, can also lead to abstinence from meat. Many families from Muslim, Hindu, but also Christian backgrounds shun pork, which is seen as an impure meat, obtained from a filthy animal. The well-known proscription of beef eating is also still very strong among Hindu families from middle and high castes. This 'taboo' has also spread towards certain Muslim communities, such as the Syed in Tamil Nadu, who consider beef as a low-status food. It also prevails within Christian families who converted from Hinduism two or three generations ago. Meanwhile, as a strategy of upward mobility, many low-caste communities have given up beef eating.

Partly built upon these multiple cultural conceptions, a political perspective also contributes to low levels of meat consumption. As a way to promote a supposedly historical Indian identity, to polarise communities and to alienate minorities (notably Muslims), political leaders and activists belonging to Hindu nationalist organisations campaign for a ban on cattle slaughter (Ahmad 2013: 122); and some of them also campaign in favour of vegetarianism. Furthermore, a new and more individual form of ethical vegetarianism is emerging among the urban middle classes. Sometimes claiming a Western influence, it intermingles moral, ritual, hygienic and sometimes environmental concerns.

Lastly, medical and dietetic regulations also account for the low intake of meat in India. Conceptions pertaining to traditional medicines establish a connection between meat eating and a higher body temperature. This 'thermal quality' (Eichinger Ferro-Luzzi 1975: 473), allegedly fostering a bad smell or a violent character, pushes many Indians to minimise their meat intake. Animal products, especially mutton, are also blamed by many Indian physicians (often high caste Hindus) and newspapers for causing metabolic diseases such as hypercholesterolemia, Type-2 diabetes and high blood pressure.

In summary, there is no obvious diet transition in contemporary India. But, in spite of all the above-mentioned restrictions and regulations, there is undoubtedly a quantitative increase in meat consumption and, notably, a qualitative change in the significations and meanings endowed to meat eating (Bruckert 2018). For the urban middle classes, the pattern of meat consumption is gradually shifting away from a ceremonial one (whereby meat is a rare treat for special occasions) and towards a more secular and banal one (whereby meat is bought and consumed as a common food). Whereas meat was previously consumed once or twice a month, it is now enjoyed up to three times a week by certain households. Individual quantities eaten are increasing, and chicken takes precedence over other types of meat. Some members of the upper class now proudly make their non-vegetarianism public, turning it into a marker of freedom and cosmopolitanism. For many young men, indulging in heavy meat consumption is a way to present their boldness and masculinity within their peer group. Subordinate communities, such as Muslims, Christians, low-caste Hindus and tribal people, try to assert their habit of meat eating as a source of cultural pride (Staples 2008). In a country where vegetarianism is seen as cultural elitism, as the oppressive and hegemonic ideology imposed by members of the high castes, eating meat also becomes an act of resistance against religious conservatism. This politicisation of vegetarianism makes India one of the only countries in the world where progressive and forward-thinking activists are usually in favour of eating more meat (Srinivasan and Rao 2015: 14).

In contemporary India, the practice of eating meat benefits from the increasingly diverse 'social and economic locations' of meat shops (Ahmad 2014: 25), from a higher purchasing power, from the global circulation of ideas and values, and from the proliferation of strictly circumscribed urban places, such as restaurants, where eating is an activity less imbued with religious and moral prescriptions than it is within the household. The long-held association of meat with power and strength (Ghassem-Fachandi 2012: 179) is thus being re-appropriated and reasserted in modern and urban contexts.

The increasing visibility and circulation of meat in the urban space actually provides more opportunities for struggle over its meanings (Robbins 1999: 419). The interviews carried out in Tamil Nadu reveal the segmented nature of meat eating in India: consumers relate to this food along the intersecting lines of religion, caste, ethnicity, revenue, gender, age, place of living (rural or urban), place of consumption (home or outside), personal agency and feelings of taste and disgust. Due to pervading Hindu beliefs and norms on the one hand, and to a growing exposure to global flows of people, goods, norms, practices and ideas on the other, India forms a stage where politicisation of vegetarianism and 'massification' of meat circuits coexist and often come into conflict.

However, if urbanisation, commodification, and global capitalism recast lifestyles and senses of belonging, they also reconfigure the networks of supply and distribution. Markets are obviously 'tied to ... the material characteristics of goods' (Horowitz, et al. 2004: 1056), and the sustainability of meat consumption in India needs to be questioned through the study of the supply chains that bring the animals and their meat from the pastures to the consumers' stomachs.

Animal Husbandry Practices: Intensification and Extensification

Within the meat circuits, the most upstream activities, namely animal husbandry practices and the procurement of feeds, are the most likely to raise concerns about ecological sustainability. In the state of Tamil Nadu, meat animal farming practices are two tier: the chicken sector can be described as being 'intensive', while ruminant keeping is more 'extensive'.

The poultry industry is vertically integrated and built upon the principles of contract farming. Big companies (such as Venkateshwara and Suguna) dominate the market and supply 'growers' with one-day-old birds, but also with technical advice, medicines and feeds (a compound of soy, maize and minerals). The meat chickens (called 'broilers') are then raised in confined animal-feeding operations (CAFOs). After forty days of fast-track breeding, the company recovers the animals and the farmer is paid according to the weight taken on by the birds during that period.

Conversely, cattle, sheep, goats and pigs are mostly raised in free-range, semi-intensive or mixed conditions. Bovine animals are often tied up to a post in the farmyard or are kept in small batches in stabling. In these cases, they are fed fresh grass, leaves, straw, corn or millet flour, rice husk, rapeseed, coconut or groundnut oilcakes. The animals are sometimes brought outside for a few hours daily to graze in the open. Goats and sheep can also be tied up in the farmyard and fed grass and crop residues (see Figure 11.1). Some are reared in large flocks, under extensive and more or less nomadic conditions, grazed on harvested fields, fallow lands, barren lands and road sides. As for pigs, they usually get their food by scavenging but can also be fed domestic waste.

None of these animals are reared for the same purpose. Female cattle and buffalo are mostly kept for milk. Traditionally, males have been used as draught animals (for pulling carts or ploughs) and all bovines were valued for their dung that provided precious manure and fuel. But machines and chemicals are steadily superseding them for these last two purposes. Therefore, almost no bovine (except some male buffalo calves) are kept primarily for meat production in India. Beef is principally a by-product of milk: the potential slaughter only happens once the animals are considered as not economically useful – namely, when females are barren and males lack the strength to work. But, for cultural and political reasons, the slaughter of cattle, and especially of cows, is very restricted in India: some states (such as Chhattisgarh) ban it, while some others (such as Tamil Nadu) allow it under certain conditions. Even if almost every bovine ends up being culled (the flesh of buffaloes is increasingly exported by industrial companies), these limits on cattle slaughter entail an ineffective management of the stock: the improper age of killing and the lack of genetic improvement have led to what has been termed a 'surplus' of unproductive cattle, especially since Independence (Simoons 1979: 475). These 200 million head of cattle are now in competition with, and progressively substituted by, a growing number of buffaloes (around 100 million), whose slaughter is less controlled.

Figure 11.1 A cow and a goat in a backyard in Tamil Nadu, 2012. Photograph by the author.

Contrary to big ruminants, goats, sheep and pigs are primarily held for their meat, for the by-products (notably the hides) they render after slaughtering, and for the profitable and easily transferable capital they embody. Small ruminants are being milked and sheared only in a few states. Unsurprisingly, their dung and, for pigs, their habit of scavenging are also valued during their lifetime. In the case of broilers, there is no secondary use: all of them are kept only for meat, merely converting feed into flesh.

The Government of India is pushing for a further intensification of meat production, to increase both the quality (in terms of hygiene and nutrient value) and the quantity of meat available for its citizens. In the chicken sector, economies of scale have already been achieved countrywide: no additional increase in productivity can sensibly be expected in the short term. Higher meat production could only be achieved by raising more chickens and, subsequently, by procuring more feed. The stocking rates of ruminants are already high, and over-grazing is deteriorating pastures. A further intensification would mean more waste to dispose of, and more feed to cultivate.

So far, the feeds are, in their vast majority, locally grown and processed. The surfaces dedicated to maize and soya bean production have expanded while the yields have increased. But the country is facing an alarming shortage of green and dry fodder (Dorin and Landy 2009: 143). Cropping and grazing surfaces cannot be extended any further, while the grains intended for animal consumption start competing with those intended for humans. Therefore, it is unlikely that a further intensification of the husbandry sector will happen in the years or decades to come. Bottlenecks in the supply chain, combined with the increasing demand for meat in urban areas, pose a severe threat to the ecological sustainability of Indian agriculture as a whole.

The 'Indian Meat Dilemma'

Procuring more meat and preserving global equilibria in local ecosystems seem to be two incompatible goals. Of course, purely materialistic views must be discarded: contrary to what Marvin Harris argued, ecological factors do not thoroughly account for the restrictions on meat consumption that prevail in India (Harris 1966). However, the country is undoubtedly facing what can be named a 'meat dilemma': solving it seems to be more complicated now than ever.

On the one hand, keeping a low level of meat consumption and production would ensure a reasonable pressure on the resources (notably land, water and fertilisers), help to prevent severe food scarcity and preserve the country's self-sufficiency in the agricultural sector by making the import of concentrate feeds unnecessary. Obviously, high population densities in some places have been made possible by the prevalence of plants in the daily diet. Lower concentration in farms would probably also let the animals live a more decent life. But keeping the meat intake voluntarily low would also fail to address the challenges posed by persistent malnutrition. Above all, this scarcity would maintain meat in its position at the two ends of the social ladder: a low-status food eaten by the lower sections of the society, and a luxury good relished by the wealthy and cosmopolitan elite. Of course, the pitfall of normative nutritional discourses has to be wisely avoided: the hackneyed narrative, relentlessly repeating the urgency to double agricultural production in order to 'feed the world', barely hides its agenda, namely the aim to secure profits and better control over agrifood markets (Weis 2015: 299). Indeed, claiming that world markets should make it possible for Indians to consume as much meat as Westerners do, is unreasonable. On the contrary, a global convergence towards a lower world average would be much more conceivable. But asserting that Indians should not be granted more than 5 kg of meat yearly due to ecological and agricultural limitations would also play into the hands of those conservative and Hindu nationalist groups who celebrate a (skewed) spiritual image of the country on the global stage and leverage the symbolic importance of meat to create boundaries (Ahmad 2014: 29).

On the other hand, higher meat consumption would unquestionably help to secure a better nutritional balance for the malnourished. Most importantly, it would enforce a 'right to food' and pave the way to an effective 'food security', defined not only by the quantity of nutrients eaten, but also by the qualitative social and cultural values that a specific food takes on within a specific group. Thereby, better access to meat in India would contribute to alleviating domination through the practice of vegetarianism, and acknowledge the role of this foodstuff, not only as a bourgeois privilege for the middle classes indulging in conspicuous consumption, but also as a powerful medium for self-respect, political emancipation and social equality. But higher meat production would inevitably be achieved through the development of animal factory farming, thereby contributing to land degradation, expanded desertification, and increased concentration of nitrate pollution and greenhouse gases (mainly methane and nitrous oxide). It would result in a higher use of fertilisers and antibiotics, contributing to an increased transmission of diseases from animals to humans, depleting the genetic diversity of livestock and putting higher pressure on global food markets in the event that imports of feed and meat are required.

To what extent can this dilemma be resolved? How is it possible to pursue an increasing meat production that would at the same time be environmentally sustainable and economically self-sufficient? The challenge is an enormous one. This chapter does not pretend to come up with ready-made or easy-to-implement solutions. Nevertheless, a few options can be quickly highlighted.

Meat production is not inherently 'eco-unfriendly'. On the contrary, pastoralists and smallholders 'help protect the environment and conserve biodiversity' (Meat Atlas 2014: 53) as their herds optimise lands unsuited for crops by converting cellulose (not directly digestible by humans) into protein and by operating a precious transfer of fertility from pastures to croplands. In India, higher meat production could be achieved in a more sustainable way through better management of the resources (Smil 2002: 628–631), opting for both a further intensification (by feeding ruminants locally grown beans, peas and alfalfa instead of soy) and a further extensification of the practices (by turning more domestic, agricultural and industrial waste into feed).

A better management of stocks would entail a more effective genetic selection, for instance by keeping the fastest growing animals as reproducers. The marketing of livestock and the production of meat could be made more efficient: shorter travel distances for live animals, better-equipped livestock markets and a proper age of slaughtering would lead to a higher output with the same (or even lower) stocking rates. But, even if bovine meat production is made more effective, for instance by fattening and slaughtering male calves made 'useless' by mechanisation, the century-old social and cultural proscription on cattle flesh is still not likely to vanish. Any extra production in the beef sector would principally result in higher exports. Rather, broilers and pigs (mono-gastric animals who better convert feed into flesh) could be favoured, while further improvements in the milk economy would lower the stocking density of bovine. Indeed, raising fewer animals for a higher milk production – in other words, partially reducing the 'surplus of cattle' – could make new resources (pastures, crops, residues, etc.) available for other meat animals. Other options, such as the improvement of aquaculture and pulses cultivation, could also result in a beneficial diversification of the diets, although they would not address the issue of the cultural politics of meat in contemporary India.

Concluding Remarks: Conceiving Meat as a Biomoral Substance

Beyond these technical (if not technocratic) and quantitative considerations, the Indian meat dilemma points out the complex, conflictual and divisive status of this peculiar food. One argument which is all too readily made by some groups is that a craving for meat denotes a patriarchal and racial mindset (Adams 2000). Rather, meat consumption reflects a million-year-old evolutionary heritage and a deep-rooted cultural preference that has been a critical aspect of nutrition in almost every society (Bulliet 2005). Thus, it cannot easily be fought against for the sole (yet indisputable) reason that it causes a deleterious effect on the environment.

Investigating seriously the sustainability of meat eating, in India as elsewhere, requires taking into account the fact that meat is not only a nitrogenous substance or a generic nutrient, but also a vector of inter-corporeality and identification, which supports organic and mental transactions and which carries matter as well as meanings. On the one hand, meat circuits participate in a long-distance circulation of physical matter, organically connecting humans with animals, but also with plants, water, soil and the atmosphere. On the other hand, meat has a strong symbolic dimension. Its perishability, and the large amount usually obtained from a single animal, encourage cooperation and sharing. Its biological similarity to human flesh, and the unavoidable act of killing that is needed to obtain it, endlessly questions its legitimacy. In this way, its production and consumption carry different beliefs and regimes of value, entailing a specific way to feel, to know and to shape specific worlds (Evans and Miele 2012: 303).

These biological and moral dimensions are ontologically intermingled, making meat a substance prone to convey a plethora of significances and to provoke contestation. The practices of producing and eating meat imply establishing a metabolic and metaphysical relation between humans and our environment, and more precisely between animals and our own animality. Accordingly, a comprehensive approach to the sustainability of meat should not downplay the sensual, visceral, ethical, aesthetic and cognitive relationship that humans maintain with this highly relational biomoral substance.

Michaël Bruckert
CIRAD (Agricultural Research Centre for Development), UMR Innovation, Montpellier, France

Notes

1. Government of India, NSSO, Household Consumption of Various Goods and Services in India, 2009–2010.
2. Source: FAO, Food Outlook, Biannual Report on Global Food Markets, May 2015, p. 49.

References

Adams, C.J. (2000) *The Sexual Politics of Meat: A Feminist-Vegetarian Critical Theory*, Bloomsbury Academic, New York.

Ahmad, Z. (2013) Marginal Occupations and Modernising Cities, *Economic & Political Weekly*, 48(32): 121–131.

Ahmad, Z. (2014) Delhi's Meatscapes: Cultural Politics of Meat in a Globalizing City, *IIM Kozhikode Society & Management Review*, 3(1): 21–31.

Arnold, D. (1994) The 'Discovery' of Malnutrition and Diet in Colonial India, *Indian Economic & Social History Review*, 31(1): 1–26.

Bruckert, M. (2018) *La chair, les hommes et les dieux: La viande en Inde*, CNRS Editions, Paris.

Bulliet, R.W. (2005) *Hunters, Herders, and Hamburgers: The Past and Future of Human–Animal Relationships*, Columbia University Press, New York.

Delgado, C.K., Rosegrant, M.W. and Meijer, S. (2001) Livestock to 2020: The Revolution Continues. Paper presented at the annual meetings of the International Agricultural Trade Research Consortium (IATRC), Auckland, New Zealand, January 18–19, 2001. Published at https://ageconsearch.umn.edu/bitstream/14560/1/cp01de01.pdf. Accessed on 26 January 2019.

Dorin, B. and Landy, F. (2009) *Agriculture and Food in India: A Half-Century Review, from Independence to Globalization*, Manohar Publishers, New Delhi.

D'Silva, J. and Webster, J. (eds) (2010) *The Meat Crisis: Developing More Sustainable Production and Consumption*, Earthscan, London and Washington, DC.

Eichinger Ferro-Luzzi, G. (1975) Temporary Female Food Avoidances in Tamilnad: Interpretations and Parallels, *East and West*, 25(3/4): 471–485.

Emel, J. and Neo, H. (2015) *Political Ecologies of Meat*, Routledge, New York and London.

Evans, A.B. and Miele, M. (2012) Between Food and Flesh: How Animals are Made to Matter (and Not Matter) within Food Consumption Practices, *Environment and Planning D: Society and Space*, 30(2): 298–314.

Fiddes, N. (1992) *Meat: A Natural Symbol*, Routledge, London and New York.

Ghassem-Fachandi, P. (2012) *Pogrom in Gujarat: Hindu Nationalism and Anti-Muslim Violence in India*, Princeton University Press, Princeton, NJ.

Harris, M. (1966) The Cultural Ecology of India's Sacred cattle, *Current Anthropology*, 7(1): 51–66.

Hayes-Conroy, A. and Hayes-Conroy, J. (eds) (2016) *Doing Nutrition Differently: Critical Approaches to Diet and Dietary Intervention*, Routledge, London and New York.

Horowitz, R., Pilcher J. and Watts S. (2004) Meat for the Multitudes: Market Culture in Paris, New York City, and Mexico City over the Long Nineteenth Century, *The American Historical Review*, 109(4): 1055–1083.

Leroy, F. and Praet, I. (2015) Meat Traditions: The Co-evolution of Humans and Meat, *Appetite*, 90: 200–211.

Meat Atlas: Facts and Figures about the Animals We Eat (2014) Heinrich-Böll-Stiftung, Berlin and Brussels.

Pingali, P. and Khwaja, Y. (2004) Globalisation of Indian Diets and the Transformation of Food Supply Systems, *ESA Working Paper 04-05*. Published at http://ageconsearch.umn.edu/bitstream/23796/1/wp040005.pdf. Accessed on 25 January 2019.

Robbins, P. (1999) Meat Matters: Cultural Politics along the Commodity Chain in India, *Cultural Geographies*, 6(4): 399–423.

Sébastia, B. (2013) L'Inde, impuissante face à la malnutrition? *Diplomatie, les grands dossiers*, 14: 32–35.

Simoons, F.J. (1979) Questions in the Sacred-Cow Controversy, *Current Anthropology*, 20(3): 467–493.

Smil, V. (2002) Eating Meat: Evolution, Patterns, and Consequences, *Population and Development Review*, 28(4): 599–639.

Srinivasan, K. and Rao, S. (2015) 'Will Eat Anything That Moves': Meat Cultures in Globalising India, *Economic and Political Weekly*, 50(39): 13–15.

Staples, J. (2008) 'Go on, just try some!': Meat and Meaning-Making among South Indian Christians, *South Asia: Journal of South Asian Studies*, 31(1): 36–55.

Weis, T. (2015) Meatification and the Madness of the Doubling Narrative, *Canadian Food Studies / La Revue canadienne des études sur l'alimentation*, 2(2): 296–303.

CHAPTER 12
EATING OUTSIDE THE HOME: FOOD PRACTICES AS A CONSEQUENCE OF ECONOMIC CRISIS IN SPAIN

Mabel Gracia-Arnaiz

Introduction

This chapter argues that among the sectors of Spanish society most affected by the economic crisis of recent years, food insecurity, understood as an economic and political phenomenon, entails not only paradoxes and tensions but also alternative strategies, giving rise both to suffering as a consequence of dietary constraints and to uncertainty as a source of empowerment. The working hypothesis is that whilst food insecurity creates uncertainty, which limits the quality of life, because food procurement requires a change of tactics, environments and interlocutors, it also encourages actions that not only minimise the impact of irregular access to food on physical and emotional health, but also transforms subjective experiences of food insecurity into points of departure for new ways of providing daily sustenance. Other ways of eating outside the home (in soup kitchens, in the street, in relatives' homes, etc.) are an example of precarious situations. In this context of uncertainty, people attempt to diversify the sources of food for their households, and even incorporate the most compromising and painful of these difficulties into their everyday habits as well as into their bodies. With the foods they buy, are given, or otherwise find, they try to create or reproduce orderly meals that are acceptable (gastronomically) by cultural standards. This study hypothesises that when food is in limited or uncertain supply, people may also engage in more disorderly (gastro-anomic) food practices, substituting cheaper and lower quality foods and also altering the distribution of food among family members.

Economic Crisis and Food Aid

Over the last ten years, the socio-economic situation in Spain has changed significantly because of the global crisis (Maluquer de Motes 2013). By way of example, the average income of Spanish households fell from 27,700 euros in early 2008 to 22,700 euros in 2014 (a decline of 18 percent), while household net worth fell from 190,400 euros to 119,400 euros in 2014 (a decline of 37 percent) (Banco de España 2017). Although some macroeconomic indicators improved in 2015 and 2016 (according to the Active Population Survey, the unemployment rate fell to 20.9 percent and Spain technically came out of recession during this period), the number of unemployed people still exceeds 4.5 million. Additionally, the quality of jobs has worsened, with more tempo-rary contracts and lower wages, which is not allowing many workers to lift themselves out of poverty (Ayllón 2015). In fact, according to the At Risk of Poverty or Social Exclusion (AROPE) index, the proportion of Spain's population at risk of social exclusion rose to 29.2 percent in 2014 – over three points more than the figure for 2010 (26 percent) – and 13.3 million people are below the poverty line, of whom 3.3 million are at risk of exclusion due to severe material shortages (Gracia-Arnaiz 2017).

The progressive impoverishment caused by economic policies based on austerity has had various consequences on the lifestyles of the groups most heavily affected by cuts in the social and health services, and consequently on their diets.[1] In terms of consumption, people with fewer resources have fewer opportunities to acquire varied and high-quality food, and less frequently. Some authors point to the decline in purchasing power as the primary expla-nation for the changes in consumption of certain products, and especially those that are cheaper and less healthy (Antentas and Vivas 2014), as well as the decline in spending on food outside the home. According to Medina et al. (2016), a comparison between the data provided by the Food Consumption Panel for Catalonia in 2008, the year the crisis began, and those for 2012, confirms the downward trend in food consumption in general (746.87 kg of food per capita in 2008 compared to 705.62 kg in 2012), with the sharpest fall registered for dairy products (although not for yoghurts or cheeses, which experienced only a moderate decline), vegetables and bread, followed by meat, fruit, potatoes and fish. Conversely, there was a slight increase in the consumption of beer and pre-prepared meals. However, caution must be exercised when interpreting the dietary patterns according to this statistical source, as these studies usually only mention per capita consumption of major food groups (e.g. fats and oils), which may contain more or less nutritious or affordable foods. On the other hand, levels of consumption of products considered very healthy, such as fruits and vegetables, changed little between 2008 and 2014 (Martin 2016).

One of the most notable impacts of the increasing insecurity is appar-ent in the difficulties that people have in meeting their basic needs, as is demonstrated by the indicators in the Survey of Living Conditions (2013): 16.9 percent of households report great difficulties in making their income last until the end of the month, an increase of 3.4 percent compared to 2012, while 27 percent say they cannot afford a meal with meat, chicken or fish every day. Meanwhile, according to the latest Fundación FOESSA (Fomento de Estudios Sociales y de Sociologia Aplicada) report (Valls and Belzunegui 2014), 16 percent of Spaniards are forced to follow a poor diet due to a lack of financial resources. People with limited resources try to cut costs, and this includes food costs. This reduction in expenditure can be achieved by minimising leftovers, increasing purchases of private label goods, or by prioritising price as the main criterion when selecting and preparing meals (Gracia-Arnaiz 2014). If these measures are insufficient, they then seek help outside the home.

Government policies related to these difficulties have been contradictory over the past five years. On the one hand, there have been cuts in social welfare benefits, with a consequent increase in poverty, while at the same time it has been necessary to design or increase mechanisms and programmes to relieve the growing food insecurity (Kramer and Gracia-Arnaiz 2015). The Food Aid Plan of the European Union (EU), co-financed by EU and government funds, is the programme that currently provides the most extensive assistance of this type in Spain. Through the implementation of this plan, 115 million kilograms or litres of food were purchased in 2015, including rice, baby food, children's cereals, powdered follow-on milk, chickpeas, beans, full cream UHT milk, olive oil, canned tuna, spaghetti, canned tomato sauce, dehydrated cream of vegetables, biscuits, canned green beans and canned fruit without added sugar. These measures have been accompanied by the programme for the collection of fruit/vegetables for distribution free of charge. While the Spanish government bought 63 million kilograms or litres of food in 2010, the figure had almost doubled by 2015.

Charities and civic organisations have been mainly responsible for the management of the food aid, which has reached people in precarious situations. Supranational institutions such as Banco de los Alimentos, the Red Cross and Caritas have undertaken initiatives to store and distribute surplus basic commodities from the agribusiness sector and from private donations. The latter two institutions have also undertaken other programmes in collaboration with local authorities, such as social discounts for purchasing food, and social canteens and supermarkets. An example of these mechanisms are the pre-payment cards that enable users to buy food in supermarkets and shops without showing the purchaser's status as a claimant and thereby avoiding stigmatisation. For example, 'Caritas brings together and works with people who are in a vulnerable situation, in order that they can recuperate their confidence and dignity and thus overcome difficulties themselves. This is one of the premises which inspires the social actions of the institution and, for the last few months, has given rise to the 'Barcelona Solidaria' card. This card is a tool that facilitates people without funds ... to receive economic help in getting their food'.[2] The government has therefore decided to address the problem of food poverty by strengthening the role of these organisations, and their institutionalisation and corporatisation. According to data provided by the Spanish Federation of Food Banks (2014),[3] the federation distributed 142,124 kilograms to 8,652 charities in 2014, which reached more than 1.5 million beneficiaries – an increase of 20 percent compared to the previous year. In November 2016 alone, 22 million kilograms were collected in the IV Food Collection,[4] which took place at retail outlets, making Spain the European country where citizens donate the most food.

However, the social and health-care dimensions of this problem are unknown. Apart from the numerous monographs on poverty and social exclusion that have been produced in recent years by various organisations and observatories (the FOESSA Foundation, Caritas, the Red Cross, etc.) no official report or specific survey on hunger, food insecurity or food aid has been produced in Spain, unlike those recently produced in some other European countries (Feeding Britain 2015 in the UK,[5] and Études ABENA in France). Even the 'Food Security Barometer' produced by the Agència Catalana de Seguretat Alimentària (Catalan Food Safety Agency) (2016) from 2008 to 2015 has not added any new variables, apart from collecting data on nutritional knowledge, habits and the perception of risk.[6] Unfortunately, we have no accurate diagnoses on the extent of this phenomenon, because in reality there is no consensus about its nature. There are also no methodological tools that directly measure this issue, and much of the available data comes from secondary sources.

What we do know is that food aid has been incorporated into a range of old and new strategies for its supply, distribution and consumption among the poorest members of the population (Gracia-Arnaiz 2014), which are very similar to those adopted in other European countries (Sutton 2014; Caplan 2016). These are changes in location for purchases, the type and frequency of the products purchased and the brands chosen, with the cheapest sought out in order to reduce spending; changes when dining, with fewer meals in restaurants and bars, or celebrations at home; changes in the ways meals are prepared, avoiding the preparation of dishes that require a great deal of energy, and particularly in the amount and treatment of leftovers, in search of an equivalence between what is purchased and what is consumed, in order to minimise waste and recycle leftovers. There have also been changes in the way that food is perceived; as it is now considered a resource to which access is limited and uncertain, it must be rationalised, protected and redistributed. The unavoidable – although flexible – need to eat to survive means that all possible measures are considered.

The data in the triennial report by the Bank of Spain (Ministerio de Economía, Industria y Competividad 2017) confirm that young people – both the employed and the unemployed – are the group whose incomes and wealth have suffered most during the crisis, while the incomes and wealth of retired people have stabilised due to lower inflation rates. However, most of these pensions amount to less than one thousand euros a month. Although the subjects are heterogeneous in terms of their origin, histories and expectations, our study highlights substantial changes in eating practices in general, the way that food is obtained shows the best evidence of the tensions caused by the difficulties that have emerged. Apart from what is available in stores, people with fewer resources obtain food by means of social assistance (public canteens, grants, pre-payment cards, food stores), collecting leftovers from restaurants, supermarkets, rubbish bins and skips, handouts of food in the street, cultivating allotments, begging and even shoplifting in stores or markets. For example, eating leftovers or waste 'from the street' is a particular practice related both to economic insecurity and to the awareness that edibility can be found far beyond the conventionally 'safe' spaces of shops that sell food (Gracia-Arnaiz 2017).

Resources, Spaces and Networks

During this period, people who have experienced rapid declines in their economic security have seen their purchasing power reduced to the point that they are unable to access essential goods. Unemployment, the end of social welfare benefits and having to meet the costs of a mortgage are usually mentioned as the main causes that lead people to seek food support (Egbe 2015). While, on the one hand, feelings of helplessness and shame at the day-to-day pressures are repeatedly mentioned by the informants, on the other they also say that they consider themselves lucky to have found support from different places and people: 'At first you don't know which door to knock on, but once you've left your embarrassment behind, you find out about what's available and you start to ask around' (María, 45 years old). Indeed, the various government bodies have expanded or instituted different types of aid, which act as a type of 'moral safety valve' (Poppendieck 1998). These are used to meet the nutritional needs of the most disadvantaged groups, easing the social pressure on the state and government bodies that would arise if public opinion felt that the country's citizens included hungry people. These institutions are presented as a moral achievement, as they put into practice the values of solidarity and altruism of thousands of donors and distributors of food, and the volunteers who take part in them. However, this is the triumph of a vertical type of charity and non-judgemental solidarity (Riches and Silvasti 2014), because something that is donated or given away is not intended to change the causes of poverty, but instead to alleviate the effects. An example of this is the role played by food banks, which are championed as institutions capable of making the best use of food surpluses, and according to the food banks themselves, they prevent waste, feed the most disadvantaged groups and help businesses to increase their social responsibility and solidarity-based marketing (Gascón and Montagut 2014).

Apart from these institutions' ability to put food from agribusiness and private donations into circulation and make it reach the poorest people, it should be stressed that this is one resource among many. Food itineraries do not end at charities, as individuals try to solve the problem of a lack of regular access to food by using various interlocutors and tactics. The informants explained that one of the first measures they took to spend less on food, in addition to using leftovers, was to buy private-label products, avoid impulse buys, and make simple dishes that are quick to cook. The limited food in the larders of those receiving food aid is striking, at least in relative terms, compared to the number of times that some people leave the home to eat. In cases of extreme insecurity, many meals are not eaten at home. In these cases, strategies are used which, whilst not necessarily new or unfamiliar, are a very different way of eating outside the home. They are no longer synonymous with doing so because of work or during leisure time, but because there are various resources outside the home that mean they do not have to use their kitchens, which in some cases cannot be used due to the cost of energy used for any cooking.

A significant proportion of the respondents report that they have eaten less at home since their living conditions became insecure, and they do not cook a great deal. They explain that they can no longer manage their household budget as they did when they had one or two stable salaries and were fully employed. None of them had ever thought beforehand that they were going to be in this situation; none of the mothers thought they would have to skip one or two meals a day or have to eat less to ensure that their children would have enough: 'I dish out what there is, but it gets smaller as it goes along – my children first, then my husband, and I'm last. If I'm hungry I eat more bread… and I drink more water. It swells up in your stomach and takes your mind off it' (Carmen, 49 years old). The absence of an income and low levels of stable funds, and the demands made by imponderable factors such as paying for gas, electricity, rent and household items, leads them to 'improvise' in many areas, including food: 'I went to look where I thought they could help me… I remember the shame I felt at the food collection at the school, asking my son's teacher… if I could have some' (Rosa, 37 years old).

Such people diversify their strategies in this area, which not only consist of seeking food to take home, but also eating out wherever possible. The groups that have established the most complex itineraries are families with children in their care. Those responsible for the daily diet – who are usually women – say that the non-perishable food that charities give them does not meet their needs. They usually receive pasta and rice, milk and tuna, whereas they receive fruit, vegetables, eggs and meat much less often. It should be noted that fresh foods and those requiring refrigeration are gradually being added to the stocks of food banks and humanitarian institutions.

Informants who eat outside the home do so for different reasons and in different ways. One possibility involves eating in other 'homes'. Some families, who, perhaps for cost reasons, do not have access to energy in their home (known as the 'energy poor') have access to flats managed by charities or provided by municipal councils so that they can cook and even eat there. These have a store with non-perishable food (pasta, rice, etc.), and users bring perishable foods which they have sometimes purchased, and other times obtained from food banks or by using supermarket vouchers. They are allocated three hours per week, and they organise themselves for cooking. They usually prepare simple dishes, thereby optimising the limited time available. They cook as quickly as possible, and several dishes at once: maybe pasta with pâté or tuna; white rice with tomato sauce and a fried egg; or dehydrated vegetable soups or soups of pasta/rice in winter. Canned pulses are lightly fried with garlic or stewed with potatoes or other vegetables if these are available. In some cases, it is possible to eat in the same place, although most take lunch boxes home. In one of them, cooking workshops take place every fortnight with volunteer cooks sent by the municipal council. When there are children in the household, the midday meal that was previously eaten at home is eaten in the school canteen if their school lunch is fully or partially subsidised, and is therefore also eaten outside the home. Some children also have breakfast and their mid-morning snack at school. This aid programme is for families with limited resources and is organised by parents' associations in schools in working-class neighbourhoods. The families appreciate this, because they say: 'at least my children have a good meal once a day' (Fina, 31 years old).

Another way of eating outside the home among adults consists of 'returning' to their parental home. There are many senior citizens in Spain who provide meals for their children and grandchildren; they cook using what their children obtain from donations (if they have asked for them and do not consume them at home), but above all they use what they can buy with their pensions: 'I used to complain because my pension was less than 800 euros, but now it pays for all of us to eat – the two of us, and my children and grandchildren' (María, 72 years old). The informants say that if their relationship with their parents is good, and especially if the person who has always been responsible for cooking in the household is still alive, they do not even have to ask, because the parents are the first to notice that they lack resources. Going to their parents' home 'avoids the embarrassment of having to ask for food in other places; I prefer it a thousand times to a soup kitchen' (Marta, 41 years old). But it is not only a question of shame. They report that it is more convenient and personally more comforting. In addition to not being subject to inflexible schedules, they are able to eat familiar dishes that are also usually in line with their tastes and limited budgets: 'My mother always asks me: what shall I make for you today? And she knows what we like … But her kitchen isn't a restaurant, so we adapt to her budget. The thing is we all eat hot food together with the people we love' (Marta, 41 years old).

Going to a soup kitchen is another common way of eating outside the home, but it is not available to all informants. Although the number of soup kitchens has increased in many cities since 2009, and they provide daily meals, access is usually a result of a referral from the social services. Some informants consider them to be one option among many possibilities. They usually provide one meal, either breakfast or lunch, although a night-time soup kitchen has opened in Barcelona. It can be a resource for the whole family, but in most cases they are used by elderly couples or single people. They are run by parish churches, municipal authorities or civil organisations, and have hired staff and volunteers (Kramer and Gracia-Arnaiz 2015). Lunch is prepared entirely with products from food bank donations and private donations. In addition to maintaining the three-dish structure of the meal (starter, main and dessert), they attempt to combine and prepare food using processes that replicate more or less widespread and culturally accepted recipes (noodle casserole, paella, beef and other stews, etc.). This does not necessarily mean that the expectations and wishes of the diners are met. The hospitality is perceived in dialectical terms. Feelings of gratitude for the hospitality and the possibility of a donated meal are mixed with feelings of a loss of autonomy over food choices and satisfaction of tastes. Indeed, the diners have little (or no) autonomy in terms of deciding 'what', 'how much', 'where', 'when' and 'with whom' they eat.

Finally, eating in all these places can take place simultaneously with consuming food collected in food stores, or food obtained from collections in the street by humanitarian organisations and civic groups at specific points in the neighbourhoods of the city. However, these practices have sometimes been prohibited by the local or regional government due to a lack of sufficient health-care checks. Some of the people who regularly use these food sources are those who did not have a fixed address before the crisis, but they are now increasingly frequented by people who had never previously used them. In general, asking for food from shops or collecting food distributed in the street is seen as one of many opportunities to increase their scarce resources. The informants know the times, days and places for the food distribution – where food is usually distributed at night and where it is distributed first thing in the morning, at breakfast time. A guide has been published in Barcelona, providing information about the location of these programmes, with location maps, timetables and the type of services available.[7] These places usually distribute sandwiches of Spanish omelette, cheese or ham, some fruit, hot soups, coffee and pastries.

The informants also explain that they visit bars, supermarkets, food stores, bakeries, patisseries and cafeterias in their neighbourhood to collect any food or leftovers; this consists of leftover cakes and unsold bread, tapas, cooked dishes from the set meal, packages of food with damaged packaging, and suchlike. These businesses typically put the leftover food in bags, plastic Tupperware containers or cardboard boxes. At other times, when the practice is systematic, the informants bring their own receptacles. Other initiatives related to free food distribution are the food stores or food banks self-managed by residents of the city's most deprived neighbourhoods. Their structure is assembly-based, and they meet weekly to arrange collections from shops, transport, distribution and cooking. Stores say they are unable to meet the demand for free food distribution from religious institutions, volunteers from NGOs and users: 'Nothing gets thrown away here!' (Supermarket sales staff, 2A). In general, people choose to eat this food in the street, 'depending on how hungry I am or what the weather is like' (Enric, 52 years old), and sometimes they take it home to cook and/or share with other members of the household, or simply consume it at another time of day.

In the streets of big cities, there are people who resort to leftovers out of necessity, whereas others do so because of their anti-consumerist ideology. The latter are those who in turn tend to be involved in the organisation of 'soup discos', which occasionally take over a public space, such as a square, a market place or a hospital, in order to produce a large meal with the aim to reduce food wastage. During fieldwork in 2014, we attended the Gran Dinar in Barcelona, organised in the El Raval neighbourhood by the 'Aprofitem els Aliments' group. More than 150 volunteers were involved in cooking for some four thousand people. While it is true that because of its sporadic nature it is not a regular source of food for our informants, their participation is socially relevant. These meals aim to show that other forms of eating – which are cheaper, healthier and more sustainable – are possible. They are usually organised by people who are highly critical of the global system of agribusiness, especially with regard to its waste and environmental costs. With the economic crisis, these activists have incorporated the effects of the population's progressive economic insecurity into their ideology. They advocate horizontal democracy and the need to seek alternatives to food poverty that do not reproduce welfarist logics. Some of these platforms aim to educate the people who have been made most vulnerable by the system about the need to change the rules, by inviting them to participate actively in projects that generate resources in their communities, learning to produce food (e.g. on community allotments), or by rescuing and reusing food that has been produced but which is not considered edible. In all cases, the aim is to create a response to the economic policies that are hindering people's right to food, by involving those most affected.

Conclusion

One of the consequences associated with the process of insecurity in everyday life is related to the change that has taken place in the ways of thinking about food. Among people who have suffered most from the effects of the crisis in particular, food is considered something to which access is limited and uncertain, and which has to be rationalised, protected and redistributed. The decline in their ability to obtain food independently on a regular basis has led to the reappearance of terms such as 'shortage', 'eat what you can' and 'skipping meals' in their everyday language. The results of this study show that they manage their needs by adopting a variety of strategies, both old and new, negotiating with various individuals, groups and institutions in different contexts. The most frequent alternatives include meals originating outside the home. In some cases, these are provided or prepared by friends, family, or civil society organisations; in others, food sources include restaurant leftovers or unconsumed portions from group events. As embodiments of penury and hunger, these solutions constitute experiences of social suffering; but they are also opportunities to question the role of government and the institutional measures currently being taken to ensure the right to food. With the food they buy, are given, or find in one way or another, they try to create or reproduce food that is socially and personally acceptable. Some of these solutions highlight experiences of suffering, as asking for food or searching for it among leftovers is often accompanied by feelings of shame and guilt. However, others involve an alternative response to economic policies that are calling into question the right to food in a way that has not been experienced since the transition to democracy, as they go beyond the usual forms of assistance provided by charities, and they show that other more sustainable and equitable ways of producing and redistributing food are possible.

Mabel Gracia-Arnaiz,
Universitat Rovira i Virgili, Tarragona, Spain

Notes

1. A larger version of this text is in Gracia-Arnaiz (2017). Based on a qualitative study of people who have experienced forms of precarisation as a consequence of the economic crisis in Spain, this article explores their food itineraries – the everyday practices and trajectories through which they obtain food for themselves and their families.
2. '*Cáritas acompaña y acoge a las personas en situaciones de vulnerabilidad para que puedan recuperar la confianza y la dignidad y así superar las dificultades por sí mismas. Ésta es una de las premisas que inspira la acción social de la institución y que la llevó a impulsar, hace unos meses, la tarjeta Barcelona Solidaria. Esta herramienta facilita que las personas sin recursos … puedan recibir las ayudas económicas para la alimentación.*' Published at: https://blog.caritas.barcelona/es/ayuda-a-necesidades-basicas/caritas-implementara-la-tarjeta-barcelona-solidaria-para-alimentos-a-las-familias-en-toda-la-demarcacion/. Accessed on 29 January 2019.
3. https://www.bancodealimentos.es/la-federacion-espanola-de-bancos-de-alimentos-hace-balance-de-su-colaboracion-con-la-fundacion-amancio-ortega/. Last accessed 14 February 2019.

4. *La Gran Recogida*, an action carried out by the Spanish Federation of Food Banks every year in the last weekend of November.
5. http://www.frankfield.co.uk/upload/docs/Feeding%20Britain%20100%20Days.pdf (last accessed 14 February 2019); but it should be noted that the UK does not measure food insecurity officially. The report referred to is based only on some food banks.
6. http://acsa.gencat.cat/ca/publicacions/estudis/recerca_sociologica/barometres/. Accessed on 20 January 2019.
7. That was the case for the guide published by the Comunitat de Sant'Egidio for the City of Barcelona (http://www.santegidio.org/documenti/doc_1069/Guia_ON_2016_def.pdf, last accessed 14 February 2019).

References

Agència Catalana de Seguretat Alimentària (Catalan Food Safety Agency) (2016) *Baròmetre de la seguretat alimentària a Catalunya*. Published at: http://acsa.gencat.cat/ca/Publicacions/estudis/recerca_sociologica/barometres/barometre-de-la-seguretat-alimentaria-a-catalunya-2015/. Accessed on 31 January 2019.

Antentas, J.M. and Vivas, E. (2014). Impacto de la crisis en el derecho a una alimentación sana y saludable, *Gaceta sanitaria*, 28 (Suppl. 1): 58–61.

Ayllón, S. (2015) *Infancia, pobreza y crisis económica*, 40. Obra Social La Caixa, Barcelona.

Caplan, P. (2016) Big Society or Broken Society? Food Banks in the UK, *Anthropology Today*, 30(1): 5–9.

Egbe, M.E. (2015) Human Agency at Work: The Case of Food Insecurity in Sub-Saharan African Immigrant Population in Tarragona. In *Otras Maneras de Comer*, Odela-Fundación Alicia, Barcelona, pp. 408–501.

Gascón, J. and Montagut, X. (2014) *Alimentos Desperdiciados:Un análisis del derroche alimentario desde la soberanía alimentaria*, Icaria, Barcelona.

Gracia-Arnaiz, M. (2014) Comer en tiempos de 'crisis': nuevos contextos alimentarios y de salud en España, *Salud Pública México*, 56(6): 648–653.

Gracia-Arnaiz, M. (2017) Otras formas de comer fuera de casa: itinerarios alimentarios en un contexto de precariedad. In Diaz, C. and Novo, A. (eds) *Comer fuera de casa*, Icaria, Barcelona, pp. 107–128.

Kramer, F.B. and Gracia-Arnaiz, M. (2015) Alimentarse o nutrirse en un comedor social en España: reflexiones sobre la comensalidad, *Demetra*, 10(3): 455–466.

Maluquer de Motes, J. (2013) España en el país de las maravillas: La nueva Gran Depresión de la economía española. In Llopis, E. and Maluquer de Motes, J. (eds), *España en Crisis: Las grandes depresiones económicas 1348–2012*. Pasado y Presente, Barcelona, pp. 221–246.

Martin, V. (2016) Cincuenta años de alimentación en España, *Distribución y Consumo*, 3: 66–88.

Medina, F.X., Aguilar, A. and Fornons, D. (2016) Alimentación, cultura y economía social: Los efectos de la crisis socioeconómica en la alimentación en Cataluña (España), *Sociedade e Cultura*, 1: 55–64.

Ministerio de Economía, Industria y Competividad (2017) *Banco de España: Encuesta Financiera de las Familiarias 2014*, Published at: http://transparencia.gob.es/

transparencia/transparencia_Home/index/MasInformacion/Informes-de-interes/
Sociedad_y_bienestar/Encuesta-financiera-familias-2014.html. Accessed on 20
January 2019.

Poppendieck, J. (1998) *Sweet Charity? Emergency Food and the End of Entitlement*,
Penguin Books, London.

Poulain, J.P. (2002) *Manger Aujourd'hui, Attitudes, Normes et Pratiques*, Privat, Paris.

Riches, G. and Silvasti, T. (eds) (2014) *First World Hunger Revisited*, Palgrave
Macmillan, London.

Sutton, D. (2014) *Secrets from the Greek Kitchen*, University of California Press,
California.

Valls, F. and Belzunegui, A. (2014) *La pobreza en España desde una perspectiva de
género*. Published at: http://www.foessa2014.es/informe/uploaded/documentos_tra-
bajo/15102014141447_8007.pdf. Accessed on 20 January 2019.

CHAPTER 13
FIRST STEPS IN DEVELOPING A FOOD WASTE MANAGEMENT STRATEGY IN A UK HIGHER EDUCATION INSTITUTION: THE UNIVERSITY OF LIVERPOOL CASE STUDY

Nick Doran and Iain Young

Overview

Food waste has been identified as a global problem that has significant detrimental economic, environmental and social impacts. This is particularly evident in the UK where food waste accounts for 20 percent of its carbon dioxide emissions and where the Waste and Resources Action Programme (WRAP) estimated 7.3 million tonnes of household food waste are thrown away each year (WRAP 2015a). Moreover, it has been predicted the amount of food waste will increase in the next twenty-five years due to economic and population growth (Chen et al. 2015).

Food waste management is complex and can be difficult for local authorities and large organisations to address responsibly. It is inextricably linked to food production, food distribution and consumption, as well as to the process of dealing with food waste. While there is a significant body of research on domestic food waste in the UK, as well as waste from catering (WRAP 2015c), there is also a large body of work related to and inspired by the Courtauld Commitment (WRAP 2015c), which reviews not only household food waste, but also grocery product and packaging waste from the UK grocery sector. The data available for large organisations whose first line of business is not about food is much more limited (Papargyropoulou et al. 2016). This chapter focuses on one such organisation: the University of Liverpool.

In this chapter we present the factors that could drive and influence the shape of a food waste management strategy in a UK Higher Education (HE) institution. There is little information and less guidance about what factors might be taken into consideration for the development of a food waste strategy in an HE institution. Furthermore, food waste is not specifically referred to in the University of Liverpool Environmental Management System ISO 14001, Sustainability Strategy (2009). However, the Sustainability Strategy does provide a strategy for Sustainable Waste Disposal. This work sets out to identify precedent and applicable guidance or legislation that should be incorporated into a food waste strategy for an HE institution. It also considers the possible cooperation, collaboration or other involvement of other large organisations in the local area (the Liverpool 'Knowledge Quarter') to illustrate the opportunities for economies of scale, capacity and working in partnership to realise a vision of a more sustainable future in the Merseyside region through the development of appropriate strategies.

What Is Food Waste?

There is no one universally accepted definition of food waste. Governments, scientific publications and organisations use terms such as 'food waste', 'unavoidable food waste', and 'food loss' to refer to different things. Nevertheless, FUSIONS define food waste as any food, and inedible parts of food, removed from the food supply chain to be recovered or disposed of (FUSIONS 2016a). This is largely compatible with the term as used in the EU and UK so we have adopted this definition for this work.

Food Waste in the EU and UK

FUSIONS (2016b: 18) notes that issues with food waste are concerned with all three pillars of sustainability: environmental, economic and social: 'Food waste is an issue of importance to global food security and good environmental governance, directly linked with environmental (e.g. energy, climate change, water, availability of resources), economic (e.g. resource efficiency, price volatility, increasing costs, consumption, waste management, commodity markets) and social (e.g. health, equality) impacts'. Food waste has become more visible in policy and academic debates due to its detrimental impacts, but knowledge of the drivers that give rise to food waste throughout the supply chain is still limited (Papargyropoulou et al. 2016).

Whilst the UN Sustainable Development Goals (United Nations 2015) include a specific target relating to food waste – namely, to halve the per capita global food waste at the retail and consumer levels and reduce food losses along production and supply chains, including post-harvest losses by

2030 – the UK's waste policy is primarily driven by the EU Waste Framework Directive. This directive does not include a specific target for food waste reduction. It is important to note that while Scotland has plans to introduce a target of 33 percent food waste reduction by 2025 (Priestley 2016), there are currently no mandatory food waste reduction targets across the UK. Kerry McCarthy, MP, introduced the Food Waste (Reduction) Bill to the UK Parliament on 9 September 2015 in an effort to set specific, legislative targets against food waste, particularly for supermarkets, manufacturers and distributors, but this bill did not progress any further.

It is estimated that around 89 million tonnes of food are wasted annually in the EU (Table 13.1), with associated costs estimated at 143 billion euros. Studies carried out between 2013 and 2016 estimated that post-farm-gate food chain food waste in the UK – that is to say, households, hospitality and food service, food manufacture, retail and wholesale sectors – amounted to around 10 million tonnes, 60 percent of which could have been avoided. This equals around 25 percent of the food that is purchased in the UK. By weight, household food waste makes up 70 percent of the UK post-farm-gate total, manufacturing 17 percent, hospitality and food service 9 percent and retail 2 percent (WRAP 2017a). The overall cost to the UK of food waste each year is more than £17 billion, of which around £13 billion is a cost for households (WRAP 2017a) Beyond the economic impact, food waste also has a significant negative environmental impact, as it is estimated that food waste accounts for 20 percent of the UK's carbon dioxide emissions. It also has links with social issues such as food poverty.

There are undoubtedly significant negative economic, social and environmental impacts from food waste, but while there has been some progress in the UK – for example, there was a 12 percent reduction in household food waste in the UK between 2007 and 2015, a reduction of 960,000 tonnes (WRAP 2017a) food waste remains a challenge. Information gathered from Eurostat (Eurostat 2010) shows that the UK wastes the most food when compared with other EU nations, and this is demonstrated below (Table 13.1).

Voluntary Regulation

The UK relies primarily on voluntary initiatives to tackle food waste, led primarily by WRAP (the Waste and Resources Action Programme), a registered charity supported by funding from DEFRA, the devolved administrations and, currently, the EU (Priestley 2016). It works with businesses, communities and individuals to reduce waste, develop sustainable products and use resources in an efficient way. WRAP is considered to be the leading authority and advisory body concerning food waste and associated food waste strategies and frameworks in the UK.

Table 13.1 Total food waste generation (tonnes) in EU – best estimate by member state (2006 EUROSTAT data [EWC_09_NOT_093], various national sources).

	Manufacturing	Households	Other Sectors	Total
EU27	34,755,711	37,701,761	16,820,000	89,277,472
Austria	570,544	784,570	502,000	1,858,000
Belgium	2,311,847	934,760	94,000	4,192,000
Bulgaria	358,687	288,315	27,000	674,000
Cyprus	186,917	47,819	21,000	256,000
Czech Republic	361,813	254,124	113,000	729,000
Denmark	101,646	494,914	45,000	642,000
Estonia	237,257	82,236	36,000	355,000
Finland	590,442	214,796	208,000	1,013,000
France	626,000	6,322,944	2,129,000	9,078,000
Germany	1,848,881	7,676,471	862,000	10,387,000
Greece	73,081	412,758	2,000	488,000
Hungary	1,157,419	394,952	306,000	1,858,000
Ireland	465,945	292,326	293,000	1,051,000
Italy	5,662,838	2,706,793	408,000	8,778,000
Latvia	125,635	78,983	11,000	216,000
Lithuania	222,205	111,160	248,000	581,000
Luxembourg	2,665	62,538	31,000	97,000
Malta	271	22,115	3,000	25,000
Netherlands	6,412,330	1,837,599	1,206,000	9,456,000
Poland	6,566,060	2,049,844	356,000	8,972,000
Portugal	632,395	385,063	374,000	1,391,000
Romania	487,751	696,794	1,089,000	2,274,000
Slovakia	347,773	135,854	105,000	589,000
Slovenia	42,072	72,481	65,000	179,000
Spain	2,170,910	2,136,551	3,388,000	7,696,000
Sweden	601,327	905,000	547,000	2,053,000
United Kingdom	2,591,000	8,300,000	3,500,000	14,391,000

Source: Preparatory Study on Food Waste across EU 27, European Commission (Eurostat 2010).

The three main WRAP initiatives targeting food waste reduction in the UK are the Courtauld 2025 Commitment (WRAP 2015a), the 'Love Food Hate Waste' campaign (WRAP 2013) and the Hospitality and Food Service Agreement (WRAP 2015b). 'Love Food Hate Waste' was a consumer-focused campaign. As this research concerns food waste management from an organisational perspective, we shall concentrate on the Courtauld 2025 Commitment and the Hospitality and Food Service Agreement (HaFSA).

The Courtauld 2025 Commitment is a ten-year agreement that was launched in 2015. Its overarching strategy was to bring together organisations across the food system – from producer to consumer – to make food and drink production and consumption more sustainable. The aim of this agreement was to 'cut the waste and greenhouse gas emissions associated with food and drink by at least one-fifth per person in ten years, and improve water stewardship, with cumulative savings of around £20 billion' (WRAP 2015c). The three targeted overall outcomes from 2015 to 2025, calculated as a relative reduction per head of population, were a 20 percent reduction in food and drink waste arising in the UK, a 20 percent reduction in the greenhouse gas intensity of food and drink consumed in the UK, and a reduction in impact associated with water use in the supply chain. The commitment has approximately 130 signatories, including multinational organisations such as Asda/Walmart, Coca-Cola, Heineken, KFC, Marks & Spencer, Nestlé, Unilever and Tesco. The agreement has had significant endorsement and is regarded as the leading commitment concerning the reduction of food waste in the UK.

There are signatories on the Courtauld 2025 Commitment that are important in the context of our work of examining food waste policy in an HE institution on Merseyside. These include the Environmental Association for Universities and Colleges, Merseyside Recycling and Waste Authority, and the University of Lincoln. This work will also consider whether other organisations interacting with the University are aware of and engage with the commitment, and if so, how this influences their policies.

The second voluntary initiative relevant to this research is the Hospitality and Food Service Agreement (HaFSA). This three-year agreement was launched in 2012 with the aim of reducing waste and increasing recycling rates within this sector in the UK. The agreement gained 230 signatories, covering approximately 25 percent of the UK sector – a significant endorsement. A final report (WRAP 2017b) has been released: whilst the first target concerning packaging waste was achieved (an 11 percent reduction against a 5 percent target), the target to increase the overall volume of food and packaging waste that is recycled, sent to anaerobic digestion or composted to 70 percent was not met, but fell short at 56 percent.

Household Food Waste on Merseyside

The Merseyside Recycling and Waste Authority (MRWA) was established in 1986 following the abolition of the Merseyside County Council, and has responsibility in the Merseyside area for a joint waste disposal authority, managing municipal waste collected by the five constituent councils of Merseyside or delivered to one of the authority's Household Waste Recycling

Centres. MRWA also collects information and data regarding waste composition and waste analysis to inform any waste management strategy operating in the Merseyside area. Initial indicators show that significant amounts of food waste are ending up in the residual waste bin, and that there is scope to improve recycling rates by increasing the amount of recycled food waste (MRWA 2016). Food waste amounts to 43 percent (approximately 52,000 tonnes) of non-recycled household waste – a very large percentage for residual waste.

A proposed intervention that may increase the percentage of food recycled would be to introduce a dedicated food waste stream to the kerbside collections. However, it was reported (Goulding 2017) that the reason Liverpool City Council decided not to integrate food waste collections was due to a lack of regional infrastructure for processing the material. The council noted that the cost of food waste treatment was high and, with a limited infrastructure available regionally, there was an imbalance between supply and demand, driving up costs. The same article reported that, in 2015/16, Liverpool City Council achieved a 29.2 percent recycling rate. However, over 37,000 tonnes of waste currently reaching residual bins needs to be recycled annually if the council is to meet the UK's 50 percent target by 2020. It appears that capacity and infrastructure are key barriers to increasing household food waste recycling. Research by Nick Doran at the University of Liverpool will explore if these same barriers contribute to limited food waste collections in large organisations in the Liverpool Knowledge Quarter.

Food Waste in the UK Higher Education Sector

Large universities have many of the same characteristics as a small town: they have energy, water, waste and resource needs, and most are places where people live, eat, sleep and conduct a range of activities day and night throughout the year, and as such they require similar provision of the necessities required by a small town. Gallardo et al. (2015) highlight that the effective management of the waste produced daily by these communities should be included as one of these 'necessities'. Furthermore, many universities minimise this impact by applying several measures, including reduction of waste. Food waste, from cafeterias, halls of residence, restaurants and food outlets, as well as general office spaces, must be accounted for in this waste management process.

The Environmental Association for Universities and Colleges (EAUC) guides the higher education sector in matters concerning environmental management and has a vision to lead and empower the post-16 education sector and to make sustainability 'just good business' (EAUC 2016a). EAUC created a guide for universities and colleges to implement appropriate waste management processes. This guide does not attempt to provide comprehensive advice on managing every possible waste but aims to provide a framework for building a waste management system (EAUC 2016b). However, the guidance is limited to generalities and does not seem to incorporate industry standards nor benchmark current activity relating to waste, particularly food waste. The only frameworks and legislation cited by the EAUC in terms of waste management are those they transcribe from European waste legislation:

- framework legislation (e.g. defining waste, permitting requirements and infrastructure)
- technical standards for the operation of waste facilities to ensure a high standard of environmental protection
- requirements for specific waste streams (e.g. measures to increase recycling or to reduce hazard).

However, these are broad, lacking detail and direction, and do not specifically refer to food waste.

The Higher Education Funding Council for England (HEFCE) is a non-departmental public body in the United Kingdom, which has been responsible for the distribution of funding to universities and colleges of higher and further education in England since 1992. 'HEFCE aims to create and sustain the conditions for a world-leading system of higher education, which transforms lives, strengthens the economy, and enriches society' (HEFCE 2018). HEFCE is in the process of reviewing how it supports higher education's contribution to sustainable development through consultation. As this process is ongoing there is limited information relating to the environmental impact from universities' activities, and there is no information regarding obligations or best practice relating to waste management.

The University of Liverpool

The University of Liverpool released an institutional strategy, Strategy 2026+, restating the original mission of the University College: 'For the advancement of learning and ennoblement of life…' (University of Liverpool 2016). The Sustainability Team have adopted this idea for the 'ennoblement of life' for the purposes of the Environmental Management System ISO 14001:2015, and realised in the (unpublished) University of Liverpool Sustainability Strategy, which seeks to create an environmentally sustainable, economically feasible and socially responsible community for its stakeholders, to ensure the development of a sustainable environment through the wise use of resources and to foster the adoption of a framework considerate to environmental, social and economic factors. Sustainable waste management is integral to this strategy, which aims to develop and implement waste management practices, prioritising disposal in line with the 'waste hierarchy', to reduce waste sent to landfill/incineration through recycling and reuse, and to minimise, at source, the waste generated.

Knowledge Quarter Liverpool

Knowledge Quarter (KQ) Liverpool (Figure 13.1) brings together the city's key partners to collaborate in a creative environment and close the economic gap with London. To do this, KQ Liverpool sets out to create and promote the dynamic and innovative industries operating within the Knowledge Quarter. This area encompasses Liverpool John Moores University (LJMU), the University of Liverpool, the Liverpool School of Tropical Medicine (LSTM) and the Royal Liverpool Hospital, and there is an overarching vision for this area to become a world-class destination for science, innovation, education, technology and the creative and performing arts.

With a mission in their Terms of Reference: 'To enhance the sustainability of the Liverpool Knowledge Quarter and be a model for the wider City Region' (KQSN 2016), the Knowledge Quarter Sustainability Network was

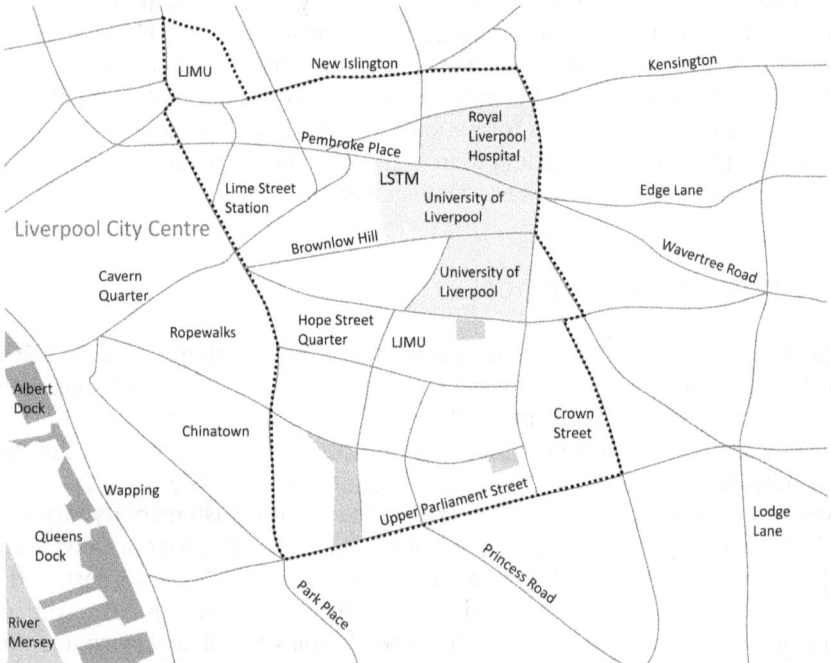

Figure 13.1 The Knowledge Quarter area in Liverpool City Centre. Figure created by author.

established in late 2014. This group meets twice a year in order to bring together key stakeholders to learn about and contribute to cross-cutting projects in the knowledge quarter that would be of mutual benefit between organisations. 'The KQSN promotes sustainability and environmental issues through Liverpool's Knowledge Quarter and encourages active collaboration and partnership. Its members comprise a vast range of experience in sustainability and are well placed across leading organisations and community groups alike to realise change and drive Liverpool's sustainability agenda' (KQSN 2016). This group may be an appropriate platform to facilitate partnership working with regards to food waste management at the University of Liverpool and throughout the Knowledge Quarter. There are direct and indirect benefits to partnership working in this way – economies of scale, resilience, identity, pooling resources. Nick Doran aims to explore this further.

Plans for Further Research – Developing the Evidence Base

University of Liverpool Waste Composition – Quantitative Analysis

The first step in the process of developing a waste management plan at a university is to know the composition, amount and distribution of the waste generated (Gallardo et al. 2015). The data collection about the university's waste composition (excluding hazardous waste) has started and includes information about the type and quantity of waste on a week-by-week (or collection-by-collection) basis. Further context will be introduced by gathering data about how this waste is currently collected and disposed of, both in terms of the kind of destination and the geographical location. One of the outputs will be to produce a 'life-cycle' schema for waste at the University of Liverpool from collection to disposal.

Comparison of Food Waste Management Strategies across Similar HE Institutions

As in any new and unproven initiative, it is important to benchmark performance against other similar organisations. Therefore, University of Liverpool's performance in food waste management will be compared to other universities. This is achievable using information and data collected through the Higher Education Statistics Agency's (HESA) Estates Management Record (EMR). The EMR is a data collection initiative that began within the sector through the Association of University Estates Directors (AUDE). It provides the higher education sector with standardised, reliable and useful property information to help managers to understand current performance, promote sharing of best practice and drive improvements (HESA 2015).

One way to identify similar universities for comparison is to compare institutions that are members of the Russell Group (of which the University of Liverpool is one). These universities are primarily research-led, share similar values and are often of a similar size and activity. Another consideration in benchmarking 'similar' universities would be to compare them within geographical regions. The N8 Research Partnership is a collaboration of the eight most research-intensive universities in the North of England: Durham, Lancaster, Leeds, Liverpool, Manchester, Newcastle, Sheffield and York. This might provide both cultural and geographical similarity. In addition to current outputs, many of the universities share information about their waste management strategy, philosophy and goals online, and are open to discussion and sharing of best practice. A University of Liverpool project is under way to examine these alongside EMR data in order to provide context and potentially causation for differing performance.

Semi-Structured Interview with Sustainability/Waste Professionals in the Knowledge Quarter

Initial research has identified several opportunities and barriers for the introduction of a bespoke food waste strategy at the University of Liverpool. For example, capacity for collections has been identified as a possible barrier to the introduction of a sustainable waste management system for food waste (Goulding 2017). To inform the development of the food waste strategy further, data will be collected about how other large organisations located in the KQ area are managing their food waste. This will also allow exploration of the possibility of taking a partnership approach to waste management. Semi-structured interviews will be carried out.

Nick Doran
University of Liverpool, UK

Iain Young
Institute of Clinical Sciences, University of Liverpool, UK

References

Chen, H., Jiang, W., Yang, Yu, Yang, Yan and Man, X. (2015) State of the Art on Food Waste Research: A Bibliometrics Study from 1997 to 2014, *Journal of Cleaner Production*, 140(2): 840–846.

EAUC (2016a) EAUC – About Us. Published at: http://www.eauc.org.uk/about_us.

Accessed on 6 February 2019.

EAUC (2016b) EAUC Waste Management Guide. Published at: http://www.sustainabilityexchange.ac.uk/waste_guide1. Accessed on 6 February 2019.

Eurostat (2010) Preparatory Study on Food Waste across EU 27, European Commission. Published at: http://ec.europa.eu/environment/archives/eussd/pdf/bio_foodwaste_report.pdf. Accessed on 6 February 2019.

FUSIONS (2016a) Food Waste Definition. Published at: https://www.eu-fusions.org/index.php/about-food-waste/280-food-waste-definition. Accessed on 6 February 2019.

FUSIONS (2016b). Estimates of European Food Waste Levels. Published at: https://www.eu-fusions.org/phocadownload/Publications/Estimates%20of%20European%20food%20waste%20levels.pdf . Accessed on 6 February 2019.

Gallardo, A., Edo-Alcón, N., Carlos, M. and Renau, M. (2015) The Determination of Waste Generation and Composition as an Essential Tool to Improve the Waste Management Plan of a University, *Waste Management*, 53: 3–11.

Goulding, T. (2017) Liverpool Opts against Food Waste Collection in Recycling Drive. Published at: http://www.letsrecycle.com/news/latest-news/liverpool-infrastructure-gap-stifles-food-waste-options/. Accessed on Accessed on 6 February 2019.

HEFCE (2018) About HEFCE. Published at: www.hefce.ac.uk/about/. Accessed on 18 April 2018.

HESA (2015) Estates Management Record. Published at: https://www.hesa.ac.uk/collection/c15042/introduction. Accessed on 6 February 2019.

KQSN (2016) Liverpool Knowledge Quarter Sustainability Network, Terms of Reference. https://www.ljmu.ac.uk/~/media/files/ljmu/about-us/sustainability/sustainability_liverpool_kqsn_terms_of_reference.pdf?la=en. Accessed on 6 February 2019.

MRWA (2016) Merseyside and Halton Waste Composition Study 2015/16. http://www.merseysidewda.gov.uk/waste-strategy/waste-analysis/. Accessed on 6 February 2019.

Papargyropoulou, E., Wright, N., Lozanoc, R., Steinberger, J., Padfield, R., na d Uljang, Z. (2016) Conceptual Framework for the Study of Food Waste Generation and Prevention in the Hospitality Sector, *Waste Management*, 49: 326–336.

Priestley, S. (2016) Food Waste Facts: Parliamentary Briefing Paper CBP07552. Published at: http://researchbriefings.files.parliament.uk/documents/CBP-7552/CBP-7552.pdf. Accessed on 06 Februray 2019.

United Nations (2015). Sustainable Development Goals. Published at: https://sustainabledevelopment.un.org/ Accessed on 6 February 2019.

University of Liverpool (2016) Strategy 2026. Published at: https://www.liverpool.ac.uk/strategy-2026/. Accessed on 6 February 2019.

WRAP (2007) Understanding Food Waste: Key findings of our recent research on the nature, scale and causes of household food waste. Published at: http://www.wrap.org.uk/sites/files/wrap/FoodWasteResearchSummaryFINALADP29_3__07.pdf. Accessed on 29 January 2019.

WRAP (2013) Love Food Hate Waste. Published at: https://www.lovefoodhatewaste.com/.

WRAP (2015a). Household Food Waste in the UK, 2015. Published at: http://www.wrap.org.uk/sites/files/wrap/Household_food_waste_in_the_UK_2015_Report.pdf. Accessed on 16 April 2018.

WRAP (2015b). Hospitality and Food Service Agreement. Published at: http://www.
    wrap.org.uk/category/sector/hospitality-and-food-service. Accessed on 9 February
    2019.
WRAP (2015c) The Courtauld Commitment 2025. Published at: http://www.wrap.org.
    uk/content/courtauld-commitment-2025. Accessed on 6 February 2019.
WRAP (2017a). Estimates of Food Surplus and Waste Arisings in the UK. Published
    at: http://www.wrap.org.uk/sites/files/wrap/Estimates_%20in_the_UK_Jan17.pdf.
    Accessed on 17 April 2018.
WRAP (2017b). The Hospitality and Food Service Agreement – Taking Action on
    Waste.   Published   at: http://www.wrap.org.uk/sites/files/wrap/Hospitality_and_
    Food_Service_Agreement_final_report_0.pdf. Accessed on 6 February 2019.

CHAPTER 14
THE DEMAND FOR SUSTAINABLE WAYS OF DEALING WITH WASTE FROM AGRICULTURE AND AQUACULTURE

Iain Young

Introduction

The agriculture and aquaculture sectors are constantly trying to improve sustainability and reduce their environmental impact. Both sectors produce a great deal of waste that is difficult to dispose of in a sustainable way. This chapter discusses the significant challenge of sustainable management of the waste produced in both the aquaculture and agriculture industries. The discussion includes an account of a recent multidisciplinary project 'BiffiO' (an acronym contrived from Biomass, Farming and Fish farming) that sets out to address this challenge by developing an economic and resource-efficient system using mixed agriculture and aquaculture waste to generate usable energy in the form of biogas and fertiliser.

Aquaculture

The production of 1 kg biomass of fish in aquaculture generates approximately 150 g of fish faeces, dry mass (Reid et al. 2008). Based on an annual fish production figure in the UK of around 200,000 tonnes (data for 2015 and 2016, Food and Agriculture Organization 2016), the UK generates almost 30,000 tonnes of fish faeces per year. In traditional flow-through and in open sea-cage fish farming, the waste products are carried out into local water courses or the sea. Waste and uneaten feed in untreated effluent or entering the sea directly from open sea cages can cause nutrient pollution, particularly in shallow or confined waterbodies (Braaten et al. 1983; Gowen and Bradbury 1987; Iwama 1991). In areas where there is intensive production, faeces, food particles and nitrogenous waste may exceed the assimilation capacity of the waters receiving them, and accumulate to levels that are toxic to marine plants and animals (Ervik et al. 1997; Hargreaves 1998).

Recirculation Aquaculture Systems (RAS) are becoming more common. RAS provides close control of culture conditions, not only improving biosecurity and facilitating the production of non-native species but also by retaining the waste produced. Retention and removal of waste is a step forward in delivering sustainable aquaculture. The aquaculture industry is also exploring the use of closed sea cages. However, while these advances offer new opportunities for the sector, they also present new challenges. The waste that was previously discharged into the environment now needs to be dealt with. The disposal of this waste represents an additional cost for the fish farms in terms of handling, processing, transport and disposal. Transport is especially an issue for many farms because aquaculture operations are often located in remote areas, with poor road networks, making the transportation of large volumes of waste difficult and costly. Furthermore, while RAS systems reduce environmental damage by collecting waste, they invariably consume more energy than the technology (e.g. sea cages) that they replace. A potential solution is the development of new technologies that can valorise the waste by generating energy and recovering valuable nutrients, thus reducing the carbon footprint of the aquaculture industry and the overall cost of these sustainable approaches.

Livestock Agriculture

It has been estimated that animal agriculture in the United States produces over a hundred times more waste than municipal sewage (Gerba and Smith 2004). For example, 2,500 head of dairy cattle produce as much waste as a town of 400,000 people (Barth 2004). Pasture-fed livestock drop their waste in the fields on which they graze, recycling nutrients into the soil. By contrast, indoor-bred livestock farming creates huge lagoons of manure that must be emptied, transported to fields and spread as fertiliser. However, manure is generally only 10–20 percent solids (or 80–90 percent water), so it is heavy and bulky, and transport costs are high.

Traditionally, in most countries with temperate climates, at least during the warmer months, livestock were grazed on pasture. Their waste dropped where they fed and the nutrients contained within it returned to the soil. Intensification of livestock farming, however, increases the numbers of 'Concentrated Animal Feeding Operations': indoor-housed or feedlot livestock, in turn producing surplus animal manure concentrated in one location. Across Europe, and indeed across much of the world, agriculture traditionally relied on the close integration of livestock and arable farming on small farms. Manure was used as the main way of enriching the soil. However, in the last few decades industrialisation and intensification of farming has led to the separation of arable and livestock farming. In Great Britain, since 1945, there has been a 65 percent decline in the number of farms and an almost fourfold increase in yield (Robinson and Sutherland 2002). In parallel, as arable agriculture intensifies, there has been an increasing demand for inorganic fertiliser. The number and extent of chemical applications in the form of fertilisers, selective herbicides and pesticides has increased greatly, and, as sources of mineral nutrients start to dwindle, prices rise. Furthermore, excessive fertilisation of arable land can damage the environment and is not economical (Goslinga and Shepherd 2005). Thus, there is a potential synergy: intensive animal production can benefit from more sustainable manure management and arable farming could potentially benefit from the redistribution of excess nutrients from manure. Intensive animal feeding operations produce manure, but the arable farms, which previously might have used this manure, are now separated geographically and, in many cases, culturally from them, with arable farmers often preferring to apply chemical fertilisers because of the comparatively low and inconsistent availability of nitrogen in livestock manure and other organic fertilisers (Feinerman and Komen 2005).

Bioenergy Production

The European Union (EU) has a goal of supplying 20 percent of its energy needs from renewable sources by 2020. It is estimated that up to 25 percent of all bioenergy in the future will come from biogas produced from wet organic materials such as animal manure, silage and feed wastes. However, there are few systems available for use by smaller fish and livestock farms. Cattle farms often have huge quantities of manure available, but manure alone has a relatively low bio-methane potential. While fish sludges seem on paper to have good bio-methane potential, the small scale of most aquaculture operations does not yield sufficient feedstock for the on-farm anaerobic digestion (AD) systems that are currently available. The BiFFiO project considered whether local livestock and fish farms could combine their waste to take advantage of the end products collectively in systems scaled to meet the demands for their specific waste needs. It was thought that mixing fish waste and manure from animal production would boost bio-methane output, thus providing a benefit to both sectors.

Biogas is a generic term used for the mixture of gases produced by the breakdown of organic matter in the absence of oxygen. It is primarily methane (CH_4) and carbon dioxide (CO_2), but frequently contains hydrogen sulphide (H_2S) and water vapour. It can be combusted as fuel to generate heat or to power an engine to generate electricity. The composition varies depending on the amount and nature of the organic solids (total solids – TS) of the sludge or manure, but there is an upper limit for TS content of around 10 percent, above which handling (mixing and pumping) becomes impossible. At the opposite end of the scale, because water content increases the amount of energy consumed to heat the slurry, slurries of 1 percent TS or less may consume the majority of biogas produced. The volatile solids (VS) content represents the fraction of the solid material that can be transformed into biogas. Most manures, including fish waste, have a VS content of 70–90 percent of TS. Anaerobic digestion (AD) typically results in a reduction of volatile solids of around 50 percent, and approximately 90 percent of the energy from the degraded organic matter ends up as methane. Biogas has a relatively low energy density (notably compared to liquid biofuels such as bioethanol) but can be purified to yield a natural gas equivalent fuel with up to 90 percent methane.

The Royal Agricultural Society of England (RASE) completed a comprehensive review of AD plants on UK farms in 2011.[1] The RASE report concludes that AD is only viable for 3.5 percent of UK dairy farms, based on current AD plant installation cost and 2011 Feed-in Tariffs (FIT), which have since become less favourable; furthermore, the generator size needed for UK farms varies from 17kWe (kW electrical) to 56kWe (Table 14.1) – a size range where the cost of installation per kWh (kW heat) produced is disproportionately high.

The benefits of small-scale, on-farm AD, however, extend beyond the financial. AD generates renewable energy and fertiliser, but there is also the potential for a reduction in pollution through reduced emissions and run-off from slurries and manures. AD can also improve nutrient availability, increasing

Table 14.1 Potential yield (kWe) from different size dairy herds.

Potential Output	Feedstock	Notes
500kWe	Slurry from 5,000 dairy cows housed year-round	This size of farm currently does not exist in the UK
50kWe	Slurry from 500 dairy cows housed year-round	This is a large UK dairy farm; but year-round housing may not be typical
10kWe	Slurry from 100 dairy cows housed year-round	This is a typical dairy farm size, but year-round housing is not typical
20kWe	Slurry from 100 dairy cows housed half the year, plus maize silage from 10ha	This would be a typical UK dairy farm, typically housed, supplemented with an energy crop or similar

Source: The Royal Agricultural Society of England Report 2011. See note 1.

the recycling of crop nutrients, reducing the amount of bought-in fertiliser, thus reducing overall carbon outputs. The digestate produced after AD generally has a lower potential as a pollutant and less odour than raw manure (Provenzano et al. 2014). In the case of cattle, manure includes whole grains and seeds that remain viable, while digestate contains far fewer viable weed seeds and has fewer live pathogens than manure, while retaining an excellent nutrition profile (Johansen et al. 2013). The co-product of AD is biogas: a mixture of methane (50–75 percent), carbon dioxide (25–50 percent), nitrogen (0–10 percent), hydrogen (0–1 percent), hydrogen sulphide (0–1 percent) and oxygen (0–2 percent). The calorific value is determined largely by the methane content and is generally around 20–25 MJ/m^3, or approximately 6 kWh/m^3, which equates to around 0.5 litres of heating oil.

The BiFFiO project

The BiFFiO project[2] was initiated by key players in the aquaculture, agriculture and renewable energy industries, who recognised a need to develop a solution for sustainable waste management. The project was funded by the 'Research for the Benefit of SMEs' instrument from the EU 7th Framework Programme used data collected from farmers and fish farmers, and publicly available data, to define the variety of operations, calculate their waste outputs and determine their energy needs. They analysed manures and fish-farm waste from representative livestock and fish farms to determine potential biogas yields. The two University of Liverpool farms – Wood Park and Ness Heath Farm – were used to represent livestock farms. Wood Park is a medium-sized, intensive dairy farm (>11,000 litres milk per head per annum) with approximately two hundred head of Holstein cattle; most of the cattle are housed indoors for most of the time. Ness Heath farm has a small herd of pedigree Hereford (beef) cattle that are kept outside for most of the year on pasture.

Two salmon hatcheries in Scotland were examined: Marine Harvest Lochailort and Tullich Hatchery. Marine Harvest Lochailort has been reported by local and national press to be the most advanced hatchery in the world, with a yield around 16 million fry per year; and the smaller Tullich Hatchery, belonging to the Scottish Salmon Company, produces around 4 million fry. Finally, two trout farms in England, Kilnsey Trout Farm and Chirk Trout Farm, were selected. Kilnsey Trout Farm in the Yorkshire Dales breeds trout to stock local lakes or to become produce for their own production facility and smoker, whilst Chirk Trout Farm in the Ceiriog Valley produces large numbers of rainbow trout for commercial and domestic use. These facilities represent the most prevalent on-shore aquaculture processors in the UK and, indeed, in Western Europe.

Waste total solids (TS) varied between 0.22 percent and 18.7 percent for the fish farm waste and 4 percent and 19.2 percent for the cattle waste, depending on site filtration/settling methods. Farms that could yield slurries of between 5 and 10 percent TS with suitable chemical properties were then selected for further biogas analysis in pilot-scale anaerobic digesters based on the Wood Park site.

Feasibility Case Studies

With the assistance of the British Trout Association and Scottish Salmon Producers' Organisation, we generated further detailed case studies for farms that were able to provide estimates of the amount and nature of the waste they produced. Most farms and fish farms do not currently perceive that they have a problem with disposal of slurries, which are generally spread on land, or in the case of the Danish fish farms, sent to central biogas facilities. Mort (fish dying in the system through illness or no identifiable cause) 'fallen stock' disposal is seen as a significant problem and it incurs a cost; most farmers highlighted the demand for a process that would reduce these costs. Costs for mort disposal are highly variable but generally around £50–100 (€60–120) depending on the route of disposal, whether there is local access to a disposal or processing facility (e.g. an amenable anaerobic digestion plant) and whether there is a need to collect and consolidate the waste. Fish mort is typically 'ensiled' for storage and to restrict the growth of bacteria. Citric, formic or propionic acid, or a combination of the same, are used, which together with the fish's own gut enzymes, break down the proteins and liquefy the fish. This fish silage is rich in nutrients and could potentially be used as fertiliser (Lo et al. 1993). However, generation of silage increases disposal and storage costs and requires space. The BiffiO project also considered how whole fish and fish processing waste or 'trimmings' could be disposed of. We carried out an analysis for each fish farm, and the closest scenario to the nature or mix of waste streams available was used to estimate potential gas outputs and the maximum usable electrical and thermal power that could be generated, as well as the applicable Feed-in Tariffs (FITs) for any excess power exported to the grid.

Discussion and Conclusion

Agricultural Farms

There are many reports and case studies available relating to AD using cattle slurry. The Royal Agricultural Society of England[1] published a series of case studies for on-farm AD which concluded that, even with the higher FITs available in 2011, AD plants would only be a viable option for around 3.5 percent of UK dairy farms, and furthermore that the generator size needed for the majority of UK farms falls into a size range where the cost per kW is disproportionately high. However, RASE also highlighted that the benefits of small-scale on-farm AD may extend beyond the gains from energy production. The digestate produced can be used as a fertiliser reducing bought-in chemical fertiliser costs and reducing overall greenhouse gas outputs. The nitrogen, phosphate and potassium in digestate are generally a higher proportion of the total volume than in the raw slurry, the carbon is reduced as methane is produced, the nitrogen is more bioavailable and the digestion process reduces the number of viable weed seeds and pathogens. Both in our survey and in the RASE report, farmers commented that the legislation and licensing process was a significant barrier to uptake. Furthermore, there was a general perception that, with decreasing FITs and Renewable Heat Incentives, and the removal of grant assistance, the costs of implementation were now prohibitive.

Aquaculture Operations

Very few fish farming operations in the UK produce sufficient waste for independent AD plant operation. However, several farms, particularly in the UK, Germany and Denmark, are located close to cattle, dairy, chicken and pig farms where large amounts of slurry could be sourced. Our mixtures including cattle waste and conclusions from previous studies (for example those detailed in the RASE report mentioned above) indicate that large-scale AD could be maintained without additional feedstock, albeit at a very low yield. Perhaps our most significant discovery is that freshwater aquaculture slurry yields almost twice the methane output per kg of solids compared to cattle manure, and, what is more, adding a proportion of processing waste and cuttings (or mort) could possibly raise this yield. However, the situation is not straightforward. Adding mort or other whole animal waste would require pre-treatment of this animal by-product material by pasteurisation to permit its later use as a fertiliser. This imposes an additional energy cost (raising the temperature of the animal by-product to 70°C and maintaining it for over one hour) and would decrease the net energy output.

Most farms we canvassed do not currently perceive that they have a problem with disposal of slurries, whereas mort disposal is seen as a significant problem. Anecdotally, there are more problems associated with the disposal of sludge (waste) in Norway where there is very little arable land close to the coastal fish farms in the west. Therefore, sludge disposal to land requires a long journey, crossing the coastal mountains to reach the more fertile valleys in the east.

There are certainly some aquaculture operations that would find AD as proposed useful. Furthermore, in the UK these operations (notably hatcheries) are increasing in number and size. Throughout Europe, and indeed worldwide, there are several companies that are seriously investigating and/ or developing the kind of large-scale Recirculation Aquaculture Systems that use sludge collection and dewatering techniques that would probably benefit from on-site AD.

Iain Young
Institute of Clinical Sciences, University of Liverpool, UK

Notes

1. Royal Agricultural Society of England Report, available at: https://www.fre-energy.co.uk/pdf/RASE-On-Farm-AD-Review.pdf. Last accessed 14 February 2019.
2. https://cordis.europa.eu/news/rcn/141328/en. Last accessed 14 February 2019.

References

Barth, E.F., Cicmanec, J.L., Goetz, J., Hantush, M.M., Herrmann, R.F., McCauley, P.T., McClellan, K.A., Mills, M.A., Richardson, T.L., Rock, S., Stoll, S., Sayles, G., Smith, J.E. and Wright S. (2004) *Risk Management Evaluation for Concentrated Animal Feeding Operations*, US Environmental Protection Agency, Washington, DC, EPA/600/R-04/042.

Braaten, B., Aure, J., Ervik, A. and Boge, E. (1983) Pollution Problems in Norwegian Fish Farming, *ICES Coastal Management*, 26: 11.

Ervik, A., Hansen, P.K., Aure, J., Stigebrandt A., Johannessen P. and Jahnsen T. (1997) Regulating the Local Environmental Impact of Intensive Marine Fish Farming, *Aquaculture*, 158: 85–94.

Feinerman, E. and Komen, M.H.C. (2005) The Use of Organic vs. Chemical Fertilizer with a Mineral Losses Tax: The Case of Dutch Arable Farmers, *Environ Resource Economics* 32(3): 367–388.

Food and Agriculture Organization (2016) Fishery and Aquaculture Country Profiles: The United Kingdom of Great Britain and Northern Ireland. Published at: http://www.fao.org/fishery/facp/GBR/en. Accessed on 7 February 2019.

Gerba, C.P. and Smith, J.E. (2004) Sources of Pathogenic Microorganisms and their Fate during Land Application of Wastes, *Journal of Environmental Quality*, 34(1): 42–48.

Goslinga, P. and Shepherd, M. (2005) Long-term Changes in Soil Fertility in Organic Arable Farming Systems in England, with Particular Reference to Phosphorus and Potassium, *Agriculture, Ecosystems and Environment*, 105(1–2): 425–432.

Gowen, R.J. and Bradbury, N.B. (1987) The Ecological Impact of Salmonid Farming in Coastal Waters: A Review, *Oceanography and Marine Biology: An Annual Review*, 25: 563–575.

Hargreaves, J.A. (1998) Nitrogen Biogeochemistry of Aquaculture Ponds, *Aquaculture*, 166: 181–212.

Iwama, G.K. (1991) Interactions between Aquaculture and the Environment, *Critical Reviews in Environmental Control*, 21: 177–216.

Johansen, A., Nielsen, H.B., Hansen, C.M., Andreasen, C., Carlsgart, J., Hauggard-Nielsen, H. and Roepstorff, A. (2013) Survival of Weed Seeds and Animal Parasites as Affected by Anaerobic Digestion at Meso- and Thermophilic Conditions, *Waste Management*, 33(4): 807–812.

Lo, K.V., Liao, P.H., Bullock, C. and Jones, Y. (1993) Silage Production from Salmon Farm Mortalities, *Aquacultural Engineering*, 12: 37–45.

Provenzano, M.R., Malerba, A.D. and Pezzolla, D.G.G. (2014) *Waste Management*, 34(3): 653–660.

Reid, G.K., Liutkus, M., Robinson, S.M.C., Chopin, T.R., Blair, T., Lander, T., Mullen, J., Page, F. and Moccia, R.D. (2008) A Review of the Biophysical Properties of Salmonid Feces: Implications for Aquaculture Waste Dispersal Models and Integrated Multi-trophic Aquaculture, *Aquaculture Research*, 40: 1–17.

Robinson, R.A. and Sutherland W.J. (2002) Post-war Changes in Arable Farming and Biodiversity in Great Britain, *Journal of Applied Ecology*, 39: 157–176.

INDEX

www.ingramcontent.com/pod-product-compliance
Lightning Source LLC
Chambersburg PA
CBHW070922030426
42336CB00014BA/2501